L.E. BRYAN

Professor and Head, Department of Microbiology and Infectious Diseases, The University of Calgary

Bacterial resistance and susceptibility to chemotherapeutic agents

CAMBRIDGE UNIVERSITY PRESS

Cambridge

London New York New Rochelle

Melbourne Sydney

Published by the Press Syndicate of the University of Cambridge
The Pitt Building, Trumpington Street, Cambridge CB2 1RP
32 East 57th Street, New York, NY 10022, USA
296 Beaconsfield Parade, Middle Park, Melbourne 3206, Australia.

First published 1982

Printed in Great Britain at the University Press, Cambridge

Library of Congress catalogue card number: 81–7724

British Library Cataloguing in Publication Data

Bryan, L.E.
Bacterial resistance and susceptibility to chemotherapeutic agents.

1. Bacteria, Pathogenic
2. Drug resistance in micro-organisms

I. Title

616′.014 QR177

ISBN 0 521 23039 X hard covers
ISBN 0 521 29785 0 paperback

CONTENTS

PREFACE

The outcome of treatment of bacterial infections with antibiotics is dependent upon many factors. Our capability to predict and to carry out successful therapy is modified by our inability to assess some of these factors particularly by conventional laboratory susceptibility testing. Many of the factors are related to conditions within the patient but many others are influenced by the structure, genetics and growth characteristics of the bacteria under test. My years as a physician, microbiologist and investigator have taught me that to understand the effective use of antibiotics and to improve on some of the important deficiencies of these agents, one should have an integrated overview of antibiotic activity. This view must take into account drug distribution and elimination in the patient, host tissue antagonism of antibiotic activity, toxicity, mechanisms of action and bacterial targets, drug penetration into bacteria, acquired and natural bacterial resistance mechanisms, the epidemiology of resistance, a clear view of where therapeutic problems exist among various clinically important bacteria and the weaknesses of our susceptibility testing systems.

Such an integrated view of antibiotics is not only useful to use drugs properly but it is useful to place emphasis in our diagnostic laboratories on overcoming such problems as slow answers to physicians on bacterial susceptibility, misleading test results, and uninterpretable and uninterpreted results. In investigational laboratories it points out where we should 'spend our money' to solve problems of therapeutic failure of antibiotics.

This book is intended to provide an overview of the many facets involved in antibiotic activity. Why is a bacterium susceptible or not susceptible to an antibiotic under a specific set of conditions? Why do bacteria change in susceptibility to antibiotics? An attempt has been made to review mechanisms of antibiotic action and resistance in the conventional manner isolated from bacteria but also in place in the whole bacterial cell. Around this approach I have tried to outline the principles influencing antibiotic activity in the patient and in the laboratory test circumstance and to point out those bacteria and those antibiotics that pose some of our major problems of antibiotic therapy. Finally I have considered some of the things we can do to overcome loss of susceptibility.

It is probably necessary that a single author write a book such as this which attempts an integrated overview of antibiotics. It is a book intended to be read rather than to be encyclopedic. At the same time, I hope it has been written with enough detailed information to provide a satisfying substance to the reader. Needless to say, as a single author, it is difficult to cover many aspects of antibiotics and to give due credit to all individuals who have made worthwhile contributions. It is also difficult to keep such a broad field up to date. Only time will tell how successfully I have met my aims and overcome the hazards of a single-author book.

I am grateful to my colleagues especially Harvey Rabin and Allan Godfrey at the University of Calgary for their part in the development of an environment in which to write this book. My thanks to those individuals who sent information before publication, in particular Dennis Kopecko, Tim Foster, Naomi Datta, Bob Hancock and Thalia Nicas. I am indebted to Joan Godfrey for typing the manuscript. Finally, I express my gratitude to my family for their interest, encouragement and especially their patience in waiting for all those late meals.

1

Antibiotic susceptibility and resistance – definition and detection

Criteria for the definition of susceptibility and resistance

The objective of laboratory testing of bacterial susceptibility to antibiotics is to identify microorganisms producing infections which either will or will not respond to conventional courses of antibiotic therapy. Such information can be used to direct initial therapy, evaluate therapy in progress or to develop susceptibility profiles of bacteria within a defined population (e.g. a hospital) as an aid to initial therapy. Response to therapy depends on both the course of antibiotics and on host factors including defense mechanisms and drug distribution. Laboratory testing methods, in general, do not fully take into account many of the factors which affect the susceptibility of bacteria within host tissues. These are discussed in later sections. The inability to assess 'host factors' means that laboratory susceptibility testing cannot give an absolute indication of *in vivo* antibiotic susceptibility. In order to provide a reasonably reliable prediction of the efficacy of an antibiotic for treating infections due to specific microorganisms, several types of criteria have been used to set susceptibility or resistance standards.

Clinical efficacy

Susceptibility of a bacterial isolate to an antibiotic is most reliably defined in relation to the record of clinical efficacy of that antibiotic. The initial assessment of efficacy (and toxicity) comes from experimental animal infections with a variety of selected bacteria. Prior to the introduction of an antibiotic, limited trials of human use are performed which provide further information on antibiotic effectiveness. These studies along with microbiological investigations provide data to establish toxic and therapeutic doses of antibiotics. However, to assess the relationship between susceptibility testing and efficacy accurately, many years of use of the agent in humans are required.

Due to a wide variation in clinical circumstances and manner of antibiotic use, it may be difficult to make absolute correlations between efficacy and the patterns of drug use. Thus, it is important to continue assessment of therapeutic response to antimicrobial agents and to re-evaluate susceptibility criteria over several years.

A practical approach to define bacterial susceptibility based on clinical response to antibiotics is to compare the susceptibility of an organism of known susceptibility status (i.e. a treatment response to a standard antibiotic dose) with that of the organism isolated from an infection. This approach requires the determination of susceptibilities of both organisms by identical diffusion or dilution methods. If the clinical isolate is inhibited by antibiotic concentrations equal to or less than that required to inhibit the control organism, it may be regarded as susceptible. If it is distinctly less sensitive the organism is regarded as resistant. Between these categories organisms may be regarded as moderately resistant or intermediate.

Control organisms are best selected by their susceptibility to systemic or urinary antibiotic concentrations rather than being the same species as the test organism. However, it is important on certain occasions to test control species of the same type as the test isolate. *Pseudomonas aeruginosa* susceptibility to aminoglycoside antibiotics is highly dependent on the concentration of magnesium and calcium in the growth medium. Thus, a clinical isolate should be compared to a *P. aeruginosa* control similarly affected by divalent cations. This organism is also relatively resistant to many of the antibiotics with which it is normally treated. It is difficult to detect small increases in resistance which may be of marked clinical significance unless the same species is used as a control. If a diffusion method is used and fastidious organisms are being treated, it is preferable to use an identical control species or an organism with similar growth characteristics.

This procedure can be sophisticated by the use of a variety of organisms susceptible to a range of antimicrobial concentrations obtained in different tissue compartments. Control organisms should cover urinary and vascular antibiotic concentrations, but on occasion controls for central nervous system, intraocular, bronchopulmonary and prostatic tissue concentrations are advisable. For ideal use of this approach numerous control strains are required which are susceptible to various concentrations of different antibiotics in tissues and which provide for the circumstances where a species of the same type is required.

Serum antibiotic levels

A second general method of defining susceptibility is based on a comparison between susceptibility concentrations and obtainable serum antibiotic levels. Inhibitory concentrations of the antibiotic in question are determined by one or more standard methods and related to serum levels produced by commonly used doses and routes of administration of the antibiotic. It has been common practice to utilize peak serum or mean (midpoint between highest and lowest serum concentrations) serum levels for comparison purposes. There are a host of variables affecting serum levels including dose, route of administration, rapid or slow infusion, renal function, body surface to mass ratio, etc. Due to the difficulty in precise definition of serum levels a safety factor has been added. Thus, it is often suggested the MIC (minimal inhibitory concentration) should be one-quarter or less of the mean serum level.

In the case of organisms for which the MIC is very much lower than serum concentrations, this method works well. Under these conditions tissue concentrations, which are often much lower than serum levels, usually are adequate to inhibit growth of the bacterial species. This relationship is illustrated by comparing commonly obtained MICs for *Streptococcus pyogenes* with benzylpenicillin blood levels obtained after oral or parenteral dosage with all but long acting penicillins. MIC values are usually in the range of 0.01 μg/ml and are well below serum levels. The predictive value of this approach is shown by a successful history of clinical efficacy of benzylpenicillin for *S. pyogenes* infection. However, in the case of many other antibiotics, MICs are much closer to peak or mean serum levels. An example is the susceptibility of *Pseudomonas aeruginosa* to carbenicillin. Frequently, MICs of carbenicillin for *P. aeruginosa* are only slightly below or equal to obtainable peak or mean serum levels. In recent years in some institutions there has been a tendency to suggest that bacteria are susceptible to antibiotics if the MIC does not exceed peak or mean serum levels. However, the evidence is that, in the case of *P. aeruginosa* and carbenicillin, this relationship does not hold as carbenicillin is frequently ineffectual as a single agent for systemic infections with this bacterium.

The problems in directly relating achievable serum concentrations and MICs fall mainly into two areas. Different methods used to determine MICs yield different results particularly for certain antibiotics. The relationship of a particular susceptibility method to circumstances within the host is usually not clear. A striking illustration of the problem is seen with determinations of MICs of aminoglycosides for *P. aeruginosa.* Alteration of divalent cation content of the medium can markedly change MIC values. A second concern is that infections are frequently partially or completely sequestrated to tissue compartments in the host. The level of antibiotic appearing in that tissue compartment and particularly at the site of infection is usually much below peak or mean serum levels.

The use of trough blood levels (blood levels taken immediately prior to the next dose) may represent a more direct relationship between MICs and serum concentrations for many antibiotics. Trough serum levels which are detectable and stable after several doses may represent an equilibrium established between the vascular compartment, the various tissue compartments and the excretion of the drug. Unfortunately no simple relationship exists between MICs and serum levels. Tissues may hold antibiotics by extensive protein binding and concentrations may occasionally be above trough levels. This is particularly likely to be the case if levels are based on single doses or widely spaced doses. It has been shown, for example, that penicillin may persist in tissues after its disappearance from blood. However, in these circumstances the tissue levels were less than 5% of peak serum levels.

In general if serum antibiotic level and MIC relationships are to be established it seems unwise to use peak levels. Mean or particularly trough levels are more likely to represent tissue concentration.

Concentrations of an antibiotic in the urine may be much above serum levels if the antibiotic is concentrated during renal excretion. For lower urinary tract infections, urinary concentrations determine the efficacy of many antibiotics. To establish susceptibility criteria for urine isolates, MICs should be related to urinary and not serum concentrations.

Population distribution of bacterial susceptibility

The third method of defining susceptibility criteria is based on comparison of the susceptibility level of a bacterial isolate to that of a large population group of the same species or similar species. Should members of that population appear which are less susceptible, they may be designated as relatively or absolutely resistant. An example of this type of situation has been observed with *Neisseria gonorrhoeae* and response to penicillin therapy. This organism responded successfully to penicillin over a span of many years. In this interval population members with reduced susceptibility to penicillin appeared. These organisms had susceptibility levels several times those of the most sensitive members of the group and were associated with an increased incidence of treatment failures. However, it was possible to successfully treat urethral infections with such gonococci using an increased dose of penicillin. Eventually a third population group appeared which had resistance levels several times those of the moderately resistant group. This group had a very high proportion of therapeutic failures and could be regarded as fully resistant to penicillin.

Resistance mechanisms

In the case of certain antibiotics it is possible to define resistance mechanisms. Bacterial isolates can be examined for the possession of such resistance mechanisms and, if found to possess them, can be regarded as resistant. This situation is best developed for penicillins. The demonstration of a *ß*-lactamase active on penicillin G in *Staphylococcus aureus* or active on ampicillin in strains of *Hemophilus influenzae* is a clear indication of antibiotic resistance.

Unfortunately, although much information is available on other resistance mechanisms, many of these are not readily detected. In addition, resistance may be associated with a combination or variety of resistance mechanisms. However, in spite of these reservations, the possession of, for example, β-lactamase activity under defined circumstances is a clear indication of resistance. The problem is that a lack of such a resistance mechanism may not always be associated with susceptibility.

In practise the definition of susceptibility to antimicrobial agents is frequently the result of testing and studies in each of the four areas noted above. Bacterial isolates which are highly susceptible or highly resistant to antimicrobial agents do not normally pose significant problems. Most bacterial isolates which are presented to clinical laboratories tend to fall into either of these two categories. Bacterial isolates that cause difficulties in defining susceptibility and which point out the limitations of our currently used methods are those of borderline antibiotic susceptibility. These bacteria are important causes of nosocomial infections. Organisms like *Streptococcus fecalis, Pseudomonas aeruginosa,* several species of nonfermentative bacteria, and many gram-negative enteric aerobic and anaerobic bacteria may fall into this group on occasion. For example, while a particular isolate of *E. coli* may be clearly susceptible to ampicillin in the urinary tract, susceptibility of the same strain causing a serious pneumonia in a patient with compromised host defenses will be very much less clear. This is often rationalized on the basis that the host has impaired defenses. However, these circumstances only illustrate that our methods of determining susceptibilities depend heavily on host defenses. In such conditions the response of a microorganism to an antimicrobial agent depends mainly on the amount of active antibiotic delivered to the infection site and the susceptibility of the organism under those conditions. It is these circumstances which are most challenging to susceptibility testing and which are the least well-explored set of variables.

Susceptibility testing methods

Methods for testing bacterial susceptibility to antibiotics fall into three general groups: diffusion and dilution assays and detection of resistance mechanisms.

Diffusion methods

Disc diffusion. In these procedures antibiotic from a central diffusion source, most commonly a paper disc, is allowed to diffuse into an agar medium during the growth of a microorganism. A continuous concentration gradient between the disc and the most peripheral antibiotic results. After a period of growth, examination is made for a zone of inhibition of bacterial growth.

The significance of the zone can be assessed by several types of criteria. It can be compared to zones produced by control 'susceptible' organisms tested under identical conditions. This is the basis of the comparative and Stokes methods of disc susceptibility testing. Zones of inhibition may, alternatively, be related to minimal inhibitory concentrations (MICs) of antibiotics. This procedure requires that a relationship be established prior to the use of the method. The procedure must be very highly standardized so that a reproducible relationship of zone and MIC can be obtained. In general a large number of strains have their MICs and zones of inhibition for individual antibiotics determined and related by regression lines obtained from plots of zone diameter versus $\log_2$ MIC. Zone criteria for susceptibility are usually based on antibiotic concentrations in blood or urine.

It is obvious that many potential problems exist for the zone : MIC method. These include variation of both MIC (a 'discontinuous' method) and zone diameter (a 'continuous' method) measurements, requirement for strictly standardized methodology, a poor appreciation of which antibiotic blood level (e.g. peak, trough or mean) to use as well as other problems.

The formation of a zone of inhibition of bacterial growth requires that a critical concentration of antibiotic be obtained in the medium which is adequate to keep growth below a critical observable cell density. When the antibiotic concentration in the medium is less than the critical concentration, growth occurs and a zone is established. A close relationship exists between the critical concentration and the MIC of a strain when determined under similar conditions.

The size of the zone depends at what distance from the central antibiotic source the critical concentration is formed. This distance depends on the rates of diffusion of the drug and of growth of the bacteria. Diffusion rates depend mainly on size, polarity, lipid solubility and chelating capabilities of the drug; temperature of incubation; concentration gradients; and thickness, pH, ionic strength and other components of the growth medium. Growth rates depend on the organism, temperature of incubation, growth atmosphere, pH, initial inoculum, nature of growth medium, the presence of antibiotic antagonists (e.g. thymidine with trimethoprim) etc.

In practise the following items require careful control particularly for the Kirby–Bauer and Ericcson diffusion methods.

Inoculum. The size of inoculum influences the time required before a critical cell density is obtained. Inoculum size is particularly important among bacteria producing β-lactamases especially those where the mechanism of resistance is principally a population phenomenon (see Chapter 5). Inoculum size is also critical in determining sulfonamide susceptibility due to the presence of adequate preformed *p*-aminobenzoic acid to allow enough growth to obscure a zone. Inocula may be applied by flooding (Ericcson), swab inoculation (Kirby–Bauer, comparative, Stokes methods) or agar overlay (Barry modification of Kirby–Bauer). In most cases semi-confluent to barely confluent growth are widely used as inocula.

Composition of the medium. The nature of the medium may markedly alter rates of antibiotic diffusion and bacterial growth. Additionally some medium constituents may antagonize or enhance antibiotic activity. Antagonists include: intermediates and end products of folate metabolism (particularly thymidine) inhibit trimethoprim and sulfonamides, preformed *p*-aminobenzoic acid antagonizes sulfonamide activity; divalent cations antagonize aminoglycoside cell entry and cell binding of polymyxins; divalent cations may chelate with tetracyclines to reduce tetracycline activity; increased phosphate concentration may reduce aminoglycoside activity; high salt concentration reduces aminoglycoside but increases bacitracin and fusidic acid activities; alkaline pH increases activity of aminoglycosides, macrolides, lincomycins and less frequently some other agents; acid pH increases activity of tetracyclines, fusidic acid, novobiocin and occasionally other agents; agar mainly through sulfonate and sulfonic acid residues bind positively charged drugs (aminoglycosides, polymyxins).

In general the addition of 5–10% blood or serum to medium tends to produce relatively small effects but could conceivably reduce zones for highly protein-bound drugs. For some highly protein-bound drugs this may more closely resemble *in vivo* conditions, although higher serum concentrations are needed to show much effect. The type of growth medium may alter zones in other undefined ways. Unfortunately batch to batch variation of the composition of growth media remains a problem. In the Kirby–Bauer method, the medium is specified as Mueller–Hinton (MH) and in the Ericcson method as MH or PDM antibiotic sensitivity medium. Those procedures using comparative zones can be performed with a variety of testing media as the control is grown under the same conditions.

Temperature of incubation. Temperature changes alter rates of antibiotic diffusion and bacterial growth. *Staphylococcus aureus* may not produce detectable methicillin resistance at temperatures greater than 35 °C.

Depth of medium. A reduction in thickness of agar depth changes rates of antibiotic diffusion. This is most pronounced with very thin agar layers where zones may markedly increase in size. A depth of 4 mm is recommended.

Conditions of incubation. Carbon dioxide (5–10%) in the atmosphere may decrease pH and reduce the activity of erythromycin, lincomycin and aminoglycosides. Anaerobic conditions markedly antagonize aminoglycoside activity. Obviously anaerobic growth conditions are required to test susceptibility of aero-sensitive bacteria.

Disc application and storage. Prediffusion of antibiotic prior to inoculation increases zone sizes. Pre-incubation of inoculated plates prior to addition of antibiotic decreases zone sizes. It is very important to dry plates before adding antibiotic discs to prevent antibiotic leaching from the disc into moisture on plates.

Although cups or wells may be used to store the antibiotic, paper dics are most widely used. Discs should be stored preferably at −20°C except for those in current use which can be stored at 4 °C. This is particularly important for various penicillins, cephalosporins and tetracyclines. It is essential that discs be kept dry until used by storage with a desiccant.

Disc content refers to the total amount of drug per disc. Contents of discs are standardized for each of the methods although more rigidly for the Kirby–Bauer and Ericcson methods. Control strains are used in all methods and usually involve strains of *S. aureus, E. coli* and *P. aeruginosa.* Specific strain numbers have been recommended in certain cases.

Zones of inhibition, if measured, should be read repeatedly in the same manner. Considerable variation in the size of the same zones recorded by different individuals can be readily observed. Comparison of zone sizes to those of susceptibility categories is best achieved through the use of templates or calipers. Lighting and definition of the zone margin are variables that require careful standardization. Rigorous quality control standards have been developed for the Kirby–Bauer method.

Other diffusion procedures. Other reservoirs of antibiotic may replace paper discs. These include paper strips, cups or cylinders, agar wells, antibiotic tablets and others. Although of value when carefully performed, these methods are in limited use.

Dilution methods

Dilution methods are used to determine MICs and may be used to determine minimal bactericidal concentrations (MBCs) of antibiotics. These procedures may be carried out with broth or in agar medium. The former is more readily adapted for determining MBCs.

Dilution methods are most widely performed using doubling or halving concentrations in broth or agar medium based on the unit of 1 (e.g. 0.25, 0.5, 1, 2, 4 etc.). As such, the method involves a discontinuous concentration gradient which partly accounts for the widely held view that the method is more precise than diffusion testing which involves a continuous antibiotic concentration gradient. The lowest concentration inhibiting visible growth is termed the MIC. In practise endpoints are usually more obvious with broth than agar medium. Agar medium, however, allows an appreciation of the number of resistant or partially inhibited organisms from the inoculum.

The major advantages of dilution testing are the much reduced effect of the rate of bacterial growth, a reduced influence of inoculum size, the capability to determine bactericidal concentrations and more readily observed endpoints. As noted under diffusion testing several conditions affect rates of drug diffusion and bacterial growth and, thus, modify zone sizes. These variables influence dilution testing to a much smaller degree.

Susceptibility of slowly growing organisms or organisms requiring demanding or unusual growth conditions are currently best determined by MIC methods. Testing of Mycobacteria, various streptococci, *Hemophilus*, *Neisseria*, strictly anaerobic bacteria and a variety of relatively unusual bacteria can be effectively tested in this fashion.

Dilution testing, however, is affected by some of the variables influencing diffusion testing. Inoculum size is a less significant variable but still can alter MICs or MBCs in the case of β-lactamase producing strains and when testing sulfonamide susceptibility. In general an inoculum of 10^5 to 10^6 organisms per milliliter is widely used. The components of growth media and the conditions of incubation noted to enhance or antagonize antibiotic activity for diffusion methods are also operative for dilution testing (see Chapter 1, Diffusion methods). As MICs are often used for organisms like *S. pneumoniae*, it is important to note that the pH of media which are poorly buffered and contain significant amounts of glucose can be markedly reduced during growth and modify susceptibility results.

MBCs are obtained by removing an aliquot from broth tubes showing obvious growth inhibition and plating these on antibiotic-free medium. Endpoints of less than 0.1% survival are frequently accepted as the MBC. MBCs can be determined from agar dilutions much less readily by using replica plating or by scraping the area of the inoculum with a loop and plating on antibiotic-free medium.

Agar dilution MICs have the advantage that many strains can be simultaneously tested using a multiple inoculator. The system has also been modified to use only a single or more often two or three selected concentrations for each antibiotic tested and which represent endpoints for susceptibility or resistance. This procedure gives much of the critical information obtained from a full series of antibiotic dilutions with a marked reduction in labor and materials. A similar selection of critical concentrations can be used in broth and this is done, for example, with the Autobac 1 automated method. However, as noted earlier the selection of endpoints can be difficult.

Another advantage of agar dilution is a better appreciation of the degree of bacterial survival at any drug concentration. MIC differences between agar and broth with the same medium are most influenced by the higher concentration of cations in agar. Aminoglycosides and tetracycline MICs tend to be higher in agar than broth. If free cations are to be added to agar for testing aminoglycoside susceptibility, binding of these to agar and polyphosphates can be reduced by the addition of sterile concentrated cation solutions to 60 °C agar after autoclaving. Other sources of discrepancies include the definition of endpoints (e.g. broth – no visible density; agar – less than 10 colonies), outgrowth of resistant colonies, β-lactamase production by *S. aureus,* impurities in agar as well as many poorly defined factors.

The development of micro methods and commercially available microtitre antibiotic dilutions, critical agar concentrations and dehydrated antibiotics in paper or plates has made MIC testing more available for routine laboratory use.

Detection of resistance mechanisms

The study of bacteria resistant to antibiotics has defined numerous mechanisms of resistance. One of these, β-lactamase activity, is readily tested for in clinical isolates of bacteria and is very frequently associated with some degree of resistance especially among *S. aureus, N. gonorrhoeae* and *H. influenzae*. Several methods exist for detection of β-lactamase activity in routine laboratories.

Acidimetric. Indicators which are sensitive to reduction of pH will change in color with the hydrolysis of the *β*-lactam ring and the associated release of a proton (amidase activity can also decrease pH but the change is usually much slower and more limited). Indicators frequently used include phenol red, Andrade's indicator and *N*-phenyl-l-naphthylamine.

Chromogenic substrate. Cephalosporin 87/312 (Nitrocefin) from Glaxo Research Ltd changes from light yellow to red upon hydrolysis of the *β*-lactam ring. It is a very sensitive indicator of *β*-lactamase; much more so than hydrolysis of another chromogenic substrate, cephacetrile.

Iodometric. Iodine is reduced by reaction with penicilloic acids and with products of *β*-lactam ring hydrolysis of cephalosporins. Reduced iodine is lost from a starch–iodine complex causing a decrease in the blue color of the complex.

Microbiologic methods. In this very sensitive procedure, test organisms are grown usually on membrane filters (or a volume of broken cell suspension is placed on the filter) placed on a solid medium containing 2–3 or more times the MIC of penicillin (or other *β*-lactam) of a sensitive indicator strain. Following growth, filters are removed and the medium is inoculated with the indicator strain. Growth occurs only in regions of *β*-lactam hydrolysis. This methodology is associated with a longer delay before reporting than those preceding.

All of the above procedures can be used with whole or broken cell suspensions. Usually *β*-lactamase of staphylococci, *Hemophilus* and *Neisseria* can be shown in whole cells. However, many gram-negative bacteria require broken cell suspensions to demonstrate low quantities of *β*-lactamase. Strains of staphylococci should be grown with 0.05 to 0.5 μg/ml methicillin or similar *β*-lactam (i.e. an inducer but not a good substrate) to induce full *β*-lactamase activity. Some strains of *S. aureus* producing *β*-lactamase give relatively low MICs or large zones of growth inhibition (though usually with heaped edges). Thus all clinically significant isolates should be tested for *β*-lactamase activity.

Although methods are available for testing for aminoglycoside-modifying enzymes, they are much more complex to detect in practise. In addition activity is highly substrate dependent and their use in routine laboratories is not widespread.

Testing of selected bacteria

Primary testing. The principal disadvantage of most susceptibility testing is the delay between submission of the sample and reporting of results. Most frequently therapy is underway and often a response or lack of response is obvious before results are available. The attraction of rapid β-lactamase testing is partly because of faster results. Similarly, inoculating the specimen directly to a testing medium and the use of a disc diffusion method produces rapid testing results.

In selected circumstances primary testing can be of value. If testing is restricted to specimens likely to contain a single or a predominating organism, valid information can be obtained. Preliminary screening of specimens by a gram-stained smear can give a good indication of organisms present. Samples from obviously highly contaminated sources (e.g. sputum, superficial ulcer, gut samples, throat, ears) are unlikely to provide much reliable information. Urine, aspirated pus, transtracheal aspirates, first day outgrowth of blood cultures, and similar specimens are more likely to be pure or have low contamination. A major concern of primary testing is poor control of inoculum density. If the procedure is used, tests with inappropriate inocula particularly with low-density inocula must be discarded and repeated by standard methods. This is especially of concern should samples from severely ill patients be tested as, for example, cerebrospinal fluid from meningitis. Low-density inocula are likely to lead to an overestimation of susceptibility, a critical error under the circumstances. Other objections to primary testing include the influence of detoxifying enzymes produced by contaminants which may not be in significant amounts in the infection, obscured results of slower growing organisms, many tests must be repeated and alteration of growth rates in mixed culture.

Testing of anaerobic bacteria. In general susceptibility of anaerobic bacteria has been determined by both diffusion and dilution methods. The requirement for anaerobiosis and for carbon dioxide by many of the organisms affects the activity of several antibiotics. These are the conditions that anaerobic bacteria require for growth and, thus, such effects almost surely exist *in vivo* as well. Additions to growth media such as hemin and vitamin K, needed for growth of some anaerobes, do not markedly affect susceptibility. Efforts to produce standardized tests are underway. It is important to emphasize that while standardization is of obvious value, it does not ensure the method is the best to indicate the *in vivo* response. It means that the results can be reproduced within certain limits of reliability.

Attention to the variabilities outlined for diffusion and dilution testing is needed here as for other testing methods. Some anaerobic bacteria grow slowly (e.g. actinomycetes, primary isolation of many anaerobic bacteria) and will require longer growth periods of 48 hours or more. Methods determining MIC and MBC remain preferable because of the large number of variables requiring control for growth of many anaerobes. Simplified methods based on selected breakpoint concentrations are of value (e.g. disc-broth methods). Disc diffusion methods have been described but are amenable only to rapidly growing anaerobes.

Mycobacteria. Susceptibility of mycobacterial isolates is of particular value for follow-up, re-treatment and treatment failures and for the 'atypical mycobacteria'. Pretreatment testing seems of value to provide baseline susceptibility of *Mycobacterium tuberculosis.* These procedures are best performed in a few laboratories that have the experience and numbers of organisms needing testing to provide reliable results. Testing is done in solid medium, usually Lowenstein–Jensen and Middlebrook 7H9 or 7H11. Drugs tested have been classified as first, second and third line agents. This classification is no longer fully valid as patterns of therapy of *M. tuberculosis* change. Drugs to be considered for testing include isoniazid, *para*-aminosalicylicate, streptomycin, rifampicin, ethambutol, ethionamide, cycloserine, viomycin, capreomycin and thiacetazone. Selection of drugs will depend on treatment patterns in that area.

Problems with individual bacteria. Some other bacteria pose specialized problems in susceptibility testing. Staphylococcal resistance to penicillinase-resistant β-lactams is best tested with methicillin or oxacillin at 35 °C or below or with 5% NaCl in the testing agar. Susceptibility to penicillin should be accompanied by examination for β-lactamase activity.

Hemophilus influenzae can be tested by disc diffusion or dilution methods. Careful inoculum control is required for both MIC (10^3 per spot on solid medium or 10^4/ml – broth) and diffusion methods or else false resistance particularly to ampicillin may be detected. β-lactamase testing is strongly advised for these organisms.

Susceptibility of *Neisseria gonorrhoeae* is most frequently performed by determining MICs on agar medium or by non-standardized diffusion testing. A variety of media are used and β-lactamases can be detected as previously described but strains with low-level penicillin resistance do not have such enzyme activity. If a sulfonamide or trimethoprim is to be tested, a low inoculum and a medium free of folate antagonists is needed (e.g. Diagnostic Sensitivity Test (DST) agar with 5–10% lyzed horse blood). Sulfonamide testing of *Neisseria meningitidis* should be performed to detect strains resistant to concentrations useful in treating nasopharyngeal carriers (MIC > 1–10 μg/ml sulfadiazine). Agar dilution methods have been described by Feldman and Abbott *et al.* (see Reeves *et al.* in references, pp. 25–7). Diffusion testing of susceptibility of *N. meningococci* can be determined using Mueller–Hinton medium or media containing blood and comparing zones of inhibition to those obtained with a susceptible control strain.

Susceptibility of the many other types of bacteria isolated from significant infections is best determined by broth or agar MICs. The effects of inoculum, growth rates, types of medium and gaseous environment are too pronounced to allow the use of diffusion testing in most cases.

Susceptibility testing of *Chlamydia* can be performed. It is most frequently done using cell cultures. Endpoints used are prevention of the formation of cell inclusions and loss of infectivity for subsequent cultures.

Susceptibility testing of combined antibiotics. It is useful in selected circumstances to determine if the combined effect of two drugs is indifferent (drug activity of each agent is unaffected by the other), antagonistic (activity of a drug is reduced by another drug) or synergistic (when the activity of two drugs is significantly greater than either drug alone at similar concentrations).

The interested reader is referred to the discussion of methods recently reviewed by Waterworth in *Laboratory Methods in Antimicrobial Chemotherapy* and Krogstad and Moellering in *Antibiotics in Laboratory Medicine* (see references).

Host factors influencing susceptibility

Bacterial susceptibility to antibiotics cannot be defined entirely by testing bacteria for inhibition of growth in culture. The ultimate objective of susceptibility testing is to predict the *in vivo* response of infectious agents to therapy with antibiotics. To do this effectively, factors which affect the activity of antibiotics *in vivo* must be taken into account. Three major sets of conditions influence *in vivo* susceptibility. These are: (a) the concentration of active drug within the infection site; (b) the bacterial phenotype resulting from growth in the host; (c) the capability of host defenses to prevent spread and to eradicate the infecting organism.

Concentration of active antibiotic at the infection site

The concentration of active antibiotic in a tissue site will depend on the distribution of the drug following oral or parenteral administration. Under usual circumstances, an antibiotic is absorbed to the central vascular compartment from which it is distributed and excreted. Tissues may be regarded as peripheral compartments which receive drug from the central vascular compartment. Only uncommonly is an antibiotic administered directly to peripheral tissues, for example, by intrathecal or subconjunctival injection.

An antimicrobial agent may be bound in serum by proteins, particularly albumin. A dissociation–association equilbrium results between free and bound drug. Drug leaves the central compartment by leaking between cytoplasmic cellular junctions, diffusion through cell membranes or occasionally by specific transport processes. Antibiotic bound to albumin may leak directly into tissues between cell junctions. Otherwise antibiotic enters tissues mainly from the unbound fraction in serum.

Tissue entry can be markedly influenced by the nature of cellular junctions as seen in the case of the 'blood–brain barrier'. Many drugs in the absence of inflammation (which enhances the leakiness between cytoplasmic junctions) enter the brain poorly. This is due to tight endothelial cell junctions and the close application of the foot processes of astrocytes to the capillary endothelium. Under these conditions of 'tight junctions', agents apparently diffuse directly through cells.

Diffusion of antibiotics depends mainly on molecular size, lipid solubility and the state of ionization of the drug. The last property will vary, depending on the pH and the dissociation characteristics of acidic and basic groups. Non-ionized forms of organic acids and bases have increased lipid solubility. If the pKa of an antibiotic differs from physiological pH by a unit, 90% of molecules will be ionized. Lipid solubility of antibiotics with ionizable groups will be very low under these conditions and will be unimportant to tissue entry.

Some antimicrobial agents may be specifically transported into tissues. Penicillins and cephalosporins are actually transported through renal tubular epithelial cells by means of a transport system for carboxylic and sulfonic acids and are transported out of the CSF. The efflux out of CSF and secretion by renal tubular cells can be inhibited to some extent by probenecid which blocks the transport system.

The practical significance of the preceding factors governing tissue entry is that tissue concentrations may vary significantly from that in serum. Antibiotics tend to penetrate poorly into the brain, cerebrospinal fluid, aqueous and vitreous humors of the eye and in many cases into bronchopulmonary secretions and prostatic tissue. In general antibiotic concentrations are much lower in bronchopulmonary secretions than in serum. For example, ampicillin in bronchopulmonary fluid may be 5–10% that of serum. Figures for other drugs include; carbenicillin 8–18%, gentamicin and tobramycin 4–50% (usually about 10%) and cefoxitin 20%. These values are for samples taken 2–3 hours after dose for both serum and bronchial secretions. Purulence may enhance antibiotic levels in sputum but only to a very limited extent. Entry of drugs to the fetal circulation is improved by high water solubility and low albumin binding. Amniotic concentrations of drugs like tetracycline, streptomycin, gentamicin and chloramphenicol are low.

Many antibiotics approximate serum levels in pleural, pericardial, ascitic and synovial fluids as well as in into the middle ear (in acute infections) and in the paranasal sinuses.

An entirely different situation exists for the urinary tract. Urinary concentrations of most penicillins, cephalosporins, tetracyclines, sulfonamides, aminoglycosides, polymyxins as well as nitrofurantoin, trimethoprim, vancomycin, isoniazid, 5-fluorocytosine and some other drugs are many times those of serum. The efficacy of antimicrobial agents for therapy of lower urinary tract infections depends on urinary antibiotic concentrations. Some agents are also concentrated in intra-renal tissue particularly aminoglycosides and penicillins but also sulfonamides, trimethoprim and some tetracyclines. However in pyelonephritis and in chronic renal disease, intra-renal concentrations may be less than in serum.

Certain drugs are concentrated in bile in active form, although in some cases they are deacetylated. The latter 'detoxification' process may reduce activity of some agents by two to four fold for gram-positive bacteria. Rifampicin, in particular, but also erythromycin, lincomycins, nafcillin and tetracyclines are concentrated in bile.

It is clearly an over-simplification to attempt to correlate bacterial susceptibility or resistance with serum antibiotic levels. In evaluating susceptibility determinations, attention must be paid to the site of the infection. Susceptibility of bacteria from infections of the brain, meninges, eye, prostate, bronchopulmonary surfaces and amniotic cavity should be interpreted cautiously. Drugs known to penetrate effectively into these areas should be selected. In general, this concern is perhaps more prominent during resolving and chronic infections. With acute infections, capillaries become more leaky and the barrier into many of these tissues is less severe. It should be noted that factors governing CSF and brain entry are not identical. Thus CSF concentrations do not necessarily reflect those of the brain.

Chloramphenicol is highly lipid soluble and bound poorly by albumin. It enters brain and CSF probably the most effectively of current antibiotics in wide use. Penicillin enters brain abscess cavities reasonably well but several other *β*-lactams do not (ampicillin, cloxacillin, cephalothin, cephaloridine). Fusidic acid, an antistaphylococcal agent enters brain tissue to a greater degree than cloxacillin. Metronidazole also achieves effective brain concentrations. Sulfadiazine, isoniazid, trimethoprim and 5-fluorocytosine enter CSF effectively. Enhanced concentrations in the CSF of penicillin, ampicillin, carbenicillin, methicillin, cephalothin, cephaloridine, rifampicin, tetracyclines, and lincomycins and, to a lesser extent other drugs, occur during acute meningitis.

Effective concentrations in the prostate are obtained by trimethoprim, doxycycline and to a lesser extent sulfonamides and erythromycin. Trimethoprim is widely distributed in tissues in general. Chloramphenicol achieves the highest concentrations in the aqueous humor of the eye of any drug examined after parenteral administration. Trimethoprim and sulfomethoxazole are also present in therapeutically effective concentrations after oral administration. Subconjunctival injection of aqueous preparations produces therapeutic levels of several drugs in aqueous humor including ampicillin and penicillin. Vitreous concentrations in animals are very low with all drugs.

Even if antimicrobial agents are delivered to the infection site in adequate concentrations, their action may be antagonized. In particular, cationic drugs like aminoglycosides and polymyxin derivatives are most readily antagonized. Divalent cations reduce the activity of both of these drug groups by competing for bacterial binding sites. Polyanions such as nucleic acids in purulent material, sulfonated glycoproteins of sputum, bacterial exopolysaccharides and cell membranes bind aminoglycosides firmly reducing their activity. Polymyxins are effectively bound to phospholipids of cell membranes. Low pH and anaerobiosis are important antagonists of aminoglycoside activity. Anaerobiosis reduces aminoglycoside transport (see Chapter 5) by decreasing cytochrome oxidase activity. Low pH also decreases transport of aminoglycosides probably by reducing the driving force for aminoglycoside entry (see Chapter 5).

Other antagonists of antibacterial agents include tissue binding of tetracyclines, *para*-aminobenzoic acid competition for dihydropteroate synthetase with sulfonamides and thymine or thymidine interference with trimethoprim action.

Some of the above effects are of very considerable magnitude. For example, MICs of aminoglycosides for *P. aeruginosa* can be increased by 4- to 32-fold by increasing Mg^{2+} from 0.2 to 1.0 mM and Ca^{2+} from 0.2 to 1.5 mM in testing media. The usual serum concentration of Mg^{2+} is about 1 mM. The free Ca^{2+} concentration is about 1.25 mM and total serum Ca^{2+} about 2.5 mM.

It is important to appreciate that certain drugs may be subject to marked antagonism and to take this into account in the interpretation of susceptibilities. Otherwise susceptibility, particularly of aminoglycoside and polymyxins, may be over-estimated perhaps resulting in low doses or the incorrect selection of one of these agents.

Binding of drugs to proteins in the inflammatory exudate also occurs. This can hypothetically have a favorable effect (as well unfavorable due to lower concentration of free drug), in that larger concentrations of a drug such as penicillin G may be accumulated compared to those resulting from free drug only.

Bacterial phenotype variation resulting from growth in or on the host

Bacteria growing in tissues frequently exhibit different phenotypes from those seen under cultural conditions. In general growth rates are slowed and rates of synthesis or turnover of target sites may be greatly reduced. New or additional exopolysaccharides may be formed and the structure of the cell envelope may be altered. Susceptibility to antimicrobial agents may vary between *in vitro* and *in vivo* growth circumstances mainly because of altered rates of antibiotic diffusion through the cell envelope or by reduced availability of transport sites or targets.

An example of changed susceptibility *in vivo* due to a change in targets is observed with penicillins. Penicillins bind to enzymes involved in synthesis, repair and destruction of peptidoglycan. Nongrowing cells are refractory to killing and slowly growing cells are resistant to killing. The basis of this observation is not known and may actually reflect a need for protein synthesis rather than growth.

The cell envelope, particularly of gram-negative bacteria, must be considered a dynamic structure. Its composition and nature varies depending on growth rates or the nutritional environment. One example is the production of the alginate exopolysaccharide of *P. aeruginosa* which is related to growth rate and is effectively produced when cultures are nitrogen-limited.

Susceptibility of *P. aeruginosa* to several antibiotics (polymyxin, aminoglycosides, EDTA) can be reduced by growth in magnesium-limited culture. The effects of cation limitation seem mediated through changes in the cell envelope. It has been shown that magnesium-deprived *P. aeruginosa* cells have a significantly increased amount of outer membrane protein Hl of 17000 to 18000 molecular weight. It has been suggested that this protein replaced Mg^{2+} at a site of the lipopolysaccharide which could otherwise bind cationic antibiotics.

Some antibiotics enter cells by energy-driven transport. In the case of streptomycin or gentamicin the presence of transport is dependent on the extent of electron transport to oxygen. Thus cells growing under anaerobic conditions (particularly in the absence of nitrate) are more resistant to aminoglycosides. Oxygen depletion of cultures is common for many laboratory grown inocula.

It is probable that bacterial growth limitation due to the low concentration of iron in serum occurs *in vivo* during infection. Bacteria capture iron by secretion of iron chelating compounds and by the addition of outer membrane protein receptors for such chelates. However, undoubtedly the growth of bacteria is frequently slowed by iron depletion *in vivo* in spite of these mechanisms to sequester iron. Iron-depleted bacteria thus grow more slowly and changes in the composition of the cell envelope and amounts of cytochromes and certain enzymes occur. These alterations are likely to influence antibiotic susceptibility *in vivo*.

Although it is obvious that the bacterial phenotype exhibited *in vivo* is important in determining antimicrobial susceptibility, it is much more difficult to state precisely why in most circumstances. *In vivo* growth circumstances are unlikely to be associated with a single specific change in bacterial structure or metabolic activity. Rather effects are a complex interaction of changes in the permeability of the cell-envelope and in the activity of energy-dependent transport and the target structures.

Host defenses

A wide variety of defects in host defenses against infection predispose to various types of infection. A partial list of many of these defects is given in Table 1.1. The presence of severely

Table 1.1. *Examples of compromised host defenses predisposing to infection*

Defects in mechanical barriers	
(1)	Burns
(2)	Cystic fibrosis (bronchopulmonary clearance)
(3)	Urinary tract obstruction
(4)	Foreign body
Defects in phagocytosis	
	Chronic granulomatous disease
	Chediak–Higashi syndrome
	Secondary granulocytopenia due to leukemia, lymphomas, malignancies and treatment of these disorders
Defects in cellular and humoral immune response	
	Primary humoral including Aldrich syndrome, type II dysgammaglobulinemia, selective IgA or IgG deficiency, Peyer's patch aplasia and dysplasia
	Primary cellular including Swiss-type agammaglobulinemia, ataxia-telongietasia, type I dysgammaglobulinemia, thymic aplasia and dysplasia
	Secondary affecting one or both types of immunity including Hodgkins disease, multiple myeloma, Waldenstroms macroglobulinemia, lymphoproliferative disorders
	Treatment including irradiation, alkylating agents, a variety of antimetabolites, antilymphocyte serum
	Splenectomy, sickle cell anemia
Defects in normal bacterial flora	
	Use of antibiotics

compromised host defenses markedly alters response to antimicrobial agents and is a major problem of modern hospital medicine. Bacterial infections in 'compromised' hosts are frequently due to *Staphylococcus aureus, Escherichia coli, Pseudomonas aeruginosa, Klebsiella* species, *Serratia marcescens* and several other gram-negative bacilli.

Conditions accounting for predisposition to infection and failure to prevent spread of infection normally involve a combination of deficiencies of defenses. The presence of severe granulocytopenia is particularly important. *Pseudomonas aeruginosa* and *Escherichia coli* have been especially troublesome pathogens among patients with severe granulocytopenia from leukemia, lymphomas, metastatic malignancies or the treatment of these diseases. Frequently such patients die from infections with organisms which are reported as susceptible to drugs like gentamicin or carbenicillin. In fact therapy of *P. aeruginosa* infections almost always involves the use of combined antimicrobial agents (an anti-pseudomonal aminoglycoside and penicillin). However, even with combined therapy, an appreciable death rate occurs with *P. aeruginosa* infections of patients in whom the underlying disease is not itself, rapidly fatal and who are infected with agents 'susceptible' to both drugs. Survival from such infections can, in many instances, be more directly attributed to control of the primary disease or due to granulocyte transfusion rather than to the use of antibiotics.

These circumstances illustrate that normal host defenses are almost a necessity for successful treatment of bacterial infections with antimicrobial agents. They also illustrate that in the presence of reduced host resistance many bacteria considered susceptible by *in vitro* testing act as if they are either resistant or of borderline susceptibility to the agents used. It is a strange anomaly of antimicrobial susceptibility testing procedures in wide use that they may be the least predictive under circumstances where effective antimicrobial therapy could be so important to patient survival. Under such conditions, susceptibility testing is most seriously tested and most seriously found wanting.

Interpretation of susceptibility testing results in patients with granulocytopenia, major defects of mechanical barriers and immunosuppression secondary to therapy or disease must be most cautious. Even the most sensitive bacteria will be very difficult to treat under these conditions. Full antibiotic dosage, selection of the most active agents and the use of drug combinations are normally required whatever the susceptibility report.

The major value of testing under these circumstances is to recognize those strains of a particular species that are much more resistant than the usual population of that species. This can be illustrated by the use of an example. Frequently strains of *P. aeruginosa* are reported as sensitive to both tobramycin and amikacin. Tobramycin is considerably more active per molecule against *P. aeruginosa* than amikacin among 'sensitive' strains. Thus, in severe sepsis, it would be the drug of choice, used in maximal dosage and with the most active anti-pseudomonal β-lactam. However some strains of *P. aeruginosa* may acetylate tobramycin but not amikacin. Only under the latter conditions is amikacin the drug of choice.

Studies performed in our laboratory at the University of Calgary have shown, however, that predictive testing can be developed in some cases of markedly impaired host defenses. In cystic fibrosis, if testing methods take into account sputum Ca^{2+}, Mg^{2+}, DNA, sulfonated glycoprotein and local antibiotic concentrations as well as *P. aeruginosa* mucoidy, testing predicts efficacy in 90% of cases and lack of efficacy in 75% of failures. In these cases therapy was with tobramycin (12–16 mg/kg) and ticarcillin (600 mg/kg).

Selected references

Amyes, S.G.B. and Smith J.T. (1978). Trimethoprim antagonists: effect of uridine in laboratory media. *J. Antimicrob. Chemother.* **4,** 421–9.

Barling, R.W.A. and Selkon, J.B. (1978). The penetration of antibiotics into cerebrospinal fluid and brain tissue. *J. Antimicrob. Chemother.* **4,** 203–27.

Barza, M. (1978). Factors affecting the intraocular penetration of antibiotics. *Scand. J. Infect. Dis. Suppl.* **14,** 151–9.

Braude, A.I. (1976). *Antimicrobial Drug Therapy.* Chapter 6. W.B. Saunders, Philadelphia.

Brown, M.R.W. (1977). Nutrient depletion and antibiotic susceptibility. *J. Antimicrob. Chemother.* **3,** 198–201.

Cevenini, R., Landini, M.P., Donati, M. and Rumpianesi, F. (1980). Antimicrobial drug susceptibility of 15 strains of Chlamydia trachomatis recently isolated from cases of non-gonococcal urethritis in Italy. *J. Antimicrob. Chemother.* **6,** 294–5.

Craig, W.A. and Suh, B. (1978). Theory and practical impact of binding of antimicrobials to serum proteins and tissues. *Scand. J. Infect. Dis. Suppl.* **14,** 92–9.

Garrod, L.O., Lambert, H.P. and O' Grady, F. (1973). *Antibiotics and Chemotherapy.* Churchill Livingstone, Edinburgh.

Hamilton-Miller, J.M.T. (1977). Towards greater uniformity in sensitivity testing. *J. Antimicrob. Chemother.* **3,** 385–92.

Jacobs, M.R., Mithal, Y., Robins-Browne, R.M., Gaspar, M.N. and Koornhof, H.J. (1977). Antimicrobial susceptibility testing of pneumococci: determination of Kirby–Bauer breakpoints for penicillin G, erythromycin, clindamycin, tetracycline, chloramphanicol and rifampin. *Antimicrob. Agents Chemother.* **16,** 190–7.

Kenny, M.A., Pollock, H.M. Minshew, B.H., Casillas, E. and Schoenknecht, F.D. (1980). Cation components of Mueller-Hinton agar affecting testing of *Pseudomonas aeruginosa* susceptibility to gentamicin. *Antimicrob. Agents Chemother.* **17,** 55–62.

Krasemann, C. and Hildenbrand, G. (1980). Interpretation of agar diffusion tests. *J. Antimicrob. Chemother.* **6,** 181–7.

Krogstad, D.J. and Mollering, R.C. (1980). Combinations of antibiotics, mechanisms of interaction against bacteria. In *Antibiotics in Laboratory Medicine,* ed. V. Lorian, pp. 298–341. Williams and Wilkins, Baltimore, London.

Lambert, H.P. (1978). Clinical significance of tissue penetration of antibiotics into the respiratory tract. *Scand. J. Infect. Dis. Suppl.* **14,** 262–6.

Lorian, V. (ed.) (1980). *Antibiotics in Laboratory Medicine,* Chapters 1, 2, 3, 4, 5, 6, and 8. Williams and Wilkins, Baltimore, London.

deLouvois, J. (1978). The bacteriology and chemotherapy of brain abscess. *J. Antimicrob. Chemother.* **4,** 395–413.

McCarthy, L.R., Mickelsen, P.A. and Grover–Smith, E. (1979). Antibiotic susceptibility of *Haemophilus vaginalis (Corynebacterium Vaginale)* to 21 antibiotics. *Antimicrob. Agents Chemother.* **16,** 186–9.

Naumann, P. (1978). The value of antibiotic levels in tissue and urine in the treatment of urinary tract infections. *J. Antimicrob. Chemother.* **4,** 9–18.

Nicas, T.I. and Hancock, R.E.W. (1980). Outer membrane protein Hl of *Pseudomonas aeruginosa.* Involvement in adaptive and mutational resistance to ethylenediamine tetraacetate, polymyxin B and gentamicin. *J. Bacteriol.* **143,** 872–8.

Norrby, R. (1979). Pharmacokinetic aspects of the treatment of infections in the central nervous system. *J. Antimicrob. Chemother.* **5,** 630–2.

Pennington. J.E. (1976). Kinetics of penetration and clearance of antibiotics in respiratory secretions in the lung. In *Immunologic and Infectious Reactions in the Lung,* ed. C.H. Kirpatrick and H.Y. Reynolds, p. 355. Marcel Dekker Inc., New York.

Rabin, H.R., Harley, F.L., Bryan, L.E. and Elfring, G.L. (1978). Evaluation of a high dose tobramycin and ticarcillin treatment protocol in cystic fibrosis based on improved susceptibility criteria and antibiotic pharmacokinetics. 1980. In *Perspectives in Cystic Fibrosis* ed. J. Sturgess, pp. 370–5. Canadian Cystic Fibrosis Foundation, Toronto.

Reeves, D.S., Phillips, I., Williams, J.D. and Wise, R. (eds.) (1978). *Laboratory Methods in Antimicrobial Chemotherapy.* Churchill Livingstone, Edinburgh.

Robins–Browne, R.M., Gaillard, M.C. and Koornhof, H.J. (1979). Antibiotic susceptibility testing of *Neisseria gonorrhoeae. J. Antimicrob. Chemother.* **5,** 67–72.

Schonfeld, H. (1978). *Pharmacokinetics,* vol. 25, *Antibiotics and Chemotherapy.* S. Darger, Basel.

Sutter, V.L., Barry, A.L., Wilkins, T.D. and Zabransky, R.J. (1979). Collaborative evaluation of a proposed reference dilution method of susceptibility testing of anaerobic bacteria. *Antimicrob. Agents Chemother.* **16,** 495–502.

Tally, F.P. (1978). Factors affecting antimicrobial agents in an anaerobic abscess. *J. Antimicrob. Chemother.* **4,** 299–301.

Wilkinson, P.J. and Reeves, D.S. (1979). Tissue penetration of trimethoprim and sulfonamides. *J. Antimicrob. Chemother.* **5** (suppl. B), 159–68.

2

Mechanism of action of antimicrobial agents

Targets of antimicrobial agents

Targets for antimicrobial agents are a diverse group involving various cellular functions. Many targets are biosynthetic enzymes, components of the cytoplasmic membrane involved in maintaining the internal cell milieu, portions of the systems used to transfer information from DNA and RNA or part of the complex assembly involved in protein synthesis.

The association between antimicrobial agents and targets may be difficult to reverse due to covalent or firm noncovalent binding or at the other extreme may be readily reversible. Aminoglycosides particularly streptomycin are known to bind firmly to the 30 S ribosomal subunit. In contrast binding of tetracycline, chloramphenicol and erythromycin is much less firm and the action of these drugs is much more reversible than that of streptomycin and many of the aminoglycosides. Drugs containing the β-lactam ring seen in penicillins, cephalosporins and other related drugs usually form covalent bonds with the target site. However, it is probable even these bonds are reversible to some extent. Ghuysen and colleagues have shown that some of the penicillin-sensitive enzymes degrade penicillins and cause their release from targets at a slow rate.

Inhibition of enzyme activity may occur by interaction of the antibiotic at the active site or at some other site of the protein molecule involved in regulating enzyme action. Competitive inhibition at the active site is not a particularly effective mode of action for an antibiotic. It requires that the antibiotic have an affinity for the active site many times that of the naturally occurring substrate. This is true because a naturally occurring substrate will accumulate and reach concentrations normally many times that of the antibiotic. Sulfonamides are a good example of the problem posed for competitive inhibition. These agents do not have a high enough affinity for dihydropteroate synthetase compared to the natural substrates to act as competitive inhibitors. It seems more likely they act to cause the formation of faulty 'dihydrofolate'. However, effective competitive inhibitors do exist as is illustrated by trimethoprim. Antibiotics can also act at allosteric sites to produce a false feedback type of inhibition. Hypothetically this is a more useful mechanism of inhibition as the synthesis of the end product which normally acts at the allosteric site is inhibited. Thus, the much higher affinity for the allosteric site is not necessary as the end product will not accumulate to high concentrations.

Antimicrobial agents must have some selective action on microbes relative to host cells to be effective. This selective toxicity is most efficient when a similar target does not exist in the host (the peptidoglycan portion of cell walls, de novo synthesis of dihydrofolate for most bacteria) or the antibiotic is excluded from the target (70 S ribosomes in the mitochondria). The selective toxicity of antimicrobial agents is often marginal and, for example, in the case of polymyxins may be due to a different distribution of phospholipids in the bacterial cytoplasmic membrane than in host cell membranes. In general the problems posed for antiviral agents are more difficult because viruses use host cell enzymes for replication. However, some viral specified functions are now known to occur and can serve as selected targets for antiviral agents. A somewhat similar problem is posed for antifungal agents which are eukaryotic cells and have many properties in common with host cells.

Some antimicrobial agents are able to kill bacteria at relatively low concentrations and are regarded as bactericidal. Frequently the basis of bactericidal activity is obscure. Agents causing leakage and more substantial damage to the cytoplasmic membrane are often bactericidal under usual test conditions. Irreversible inhibition of certain enzymes or cellular processes can lead to cell death. Thus agents covalently bound to their targets might lead to cell death. However penicillins which are covalently bound to target enzymes cause death in many cases by the activation of an autolytic system. Aminoglycosides are an example of the difficulty in defining the basis of cell death. Streptomycin firmly associates with ribosomes but it also disturbs membrane permeability. The actual basis of death remains unknown.

Targets associated with cell wall synthesis

β-lactams

In spite of a large body of knowledge concerning many of these agents, the precise mechanisms by which cell wall synthesis is inhibited is not known, in most cases. Targets and primary functions inhibited by this group of agents are given in Table 2.1. Although cell walls contain other structures in addition to peptidoglycan, these antibacterial agents interfere almost entirely only with the synthesis and control of this important structure.

Penicillins and cephalosporins (β-lactam antibiotics) bind to a series of proteins contained within the cytoplasmic membrane. These proteins represent penicillin-sensitive enzymes inhibited by β-lactam antibiotics. Each of the proteins may possess one or more functions including D, D-carboxypeptidase, transpeptidase and endopeptidase activities. Activities possessed depend upon the type of bacterium, the particular protein and the conditions used for isolation and assay. β-lactam target proteins will be referred to as penicillin-binding proteins (PBPs).

In *E. coli,* at least seven proteins have been detected which bind ^{14}C-benzylpenicillin G. Table 2.2 lists these and their probable functions and properties. PBPs 1B and 3 seem very important to the action of most β-lactam antibiotics. In contrast PBPs 4, 5 and 6 are apparently not primary targets. A clear exception to the importance of PBPs 1 and 3 is mecillinam and related penicillins. This compound is a 6 β-amidino penicillanic acid derivative in contrast to the usual β-acylamino derivatives of the penicillin or cephalosporin nucleus. It selectively binds to *E. coli* PBP-2 resulting in osmotically stable large ovoid cells. Mecillinam is much more active against many gram-negative bacteria than gram-positive bacteria. Certain oral streptococci are also susceptible to mecillinam. This differential activity may be due to the greater significance of the PBP-2 function (maintenance of rod shape) in those gram-negative bacteria on which it is very active. The structural gene for PBP-2 has been mapped at the *rod*A locus at 14 minutes on the *E. coli* chromosome. Mutations at the *rod*Y locus at 67 minutes cause a round cell phenotype and are mecillinam resistant.

Binding to PBP-3 and inhibition of its function results in a failure to form septa, prevents cell division and causes the formation of filamentous bacteria. This is a frequently observed initial morphological effect of many β-lactams on several types of gram-negative bacilli. It seems likely that most β-lactams inhibit cell division at or about the minimal inhibitory concentration of the bacterial strain.

Inhibition of the activity of PBP-1B can prevent cell elongation but it is probable that the function of PBP-1A must also be inhibited at similar concentrations of the β-lactam. PBP-1A can apparently act as a 'detour' enzyme, assuming the major transpeptidase activity in the cell if PBP-1B is deficient as, for example, in *pon*B mutants. Cephaloridine, unlike most β-lactams, inhibits activities of both PBPs 1A and 1B at concentrations somewhat less than that needed to inhibit PBP-3 function. Thus this antibiotic causes cell lysis without initially inhibiting cell division, an unusual effect of β-lactams on gram-negative bacilli.

The result of the above effects is that most β-lactams produce filamentous bacteria when they are used at low concentrations. If the concentration is raised, bulging of the cell envelope appears at points of cell division and eventually cell lysis occurs. Very high concentrations are associated with initial inhibition of PBP-3, 1B and 1A and cell lysis occurs before significant filament formation (see Fig. 2.1).

Efficacy or lack of efficacy of a β-lactam antibiotic for certain gram-negative bacteria depends on the affinity of the drug for particularly, PBPs 1B and 3. The penicillin PC-904 has a much higher affinity for PBP-3 than benzylpenicillin in *E. coli* and *P. aeruginosa*. These bacteria are much more susceptible to PC-904 than to benzylpenicillin. However, other factors also contribute to the greater efficacy of PC-904 including β-lactamase resistance and perhaps better penetration through the outer membrane of the cell wall.

Penicillin-binding proteins are also found in gram-positive bacteria and in gram-negative cocci. *Staphylococcus aureus* has

Table 2.1. *Inhibitors of bacterial cell wall synthesis*

	Target	Target function	Target location	Comments
(A) Inhibitors of biosynthetic enzymes				
(1) Phosphonomycin	pyruvate-UDP-*N*-acetyl glucosamine transferase	UDP-*N*-acetylglucosamine + phosphoenolpyruvate → phosphate + UDP-*N*-acetyl glucosamine-enolpyruvate. The latter is ultimately incorporated into peptidoglycan	intracellular	phosphonomycin is an analog of phosphoenolpyruvate
(2) D-cycloserine	(i) alanine racemase (ii) D-alanine-D-alanine synthetase	(i) L-alanine → D-alanine (ii) 2 D-alanine → D-ala-D-ala. D-ala-D-ala is a substrate for transpeptidase and D,D-carboxypeptidase	intracellular	D-cycloserine is a competitive inhibitor of both enzymes having greater affinity for them than the natural substrate
(3) Fluoro-D-alanine	alanine racemase	L-alanine → D-alanine	intracellular	competitive inhibitor
(4) Penicillins Cephalosporins	-β-lactam binding proteins -activation of murein hydrolases secondary to inhibition of peptidoglycan synthesis in, at least, some bacteria	peptidoglycan synthesis and turnover (see also Table 2.2)	cytoplasmic membrane	transpeptidases, D,D-carboxypeptidase and endopeptidases are β-lactam sensitive enzymes. β-lactams are covalently bound to target enzymes
(5) Mecillinam	β-lactam binding protein 2 of *E. coli*	cell shape	cytoplasmic membrane	

(B) Inhibitors which combine with carrier molecules				
(1) Bacitracin	pyrophosphate -C_{55}-isoprenoid alcohol carrier	transfer of peptidoglycan subunits through the hydrophobic cytoplasmic membrane to the growing point of the peptidoglycan	cytoplasmic membranes	bacitracin prevents the dephosphorylation of the carrier which will therefore not react with UDP-*N*-acetyl muramyl pentapeptide. Also modifies protoplast permeability
(C) Inhibitors which combine with substrates				
(1) Vancomycin	acyl-D-ala-D-ala terminii of pentapeptide of *N*-acetyl-muramyl-pyrophosphoryl C_{55}-isoprenoid	peptidoglycan subunit, in transit to growing point of peptidoglycan	cytoplasmic membrane	blocks transfer of subunit to growing point of peptidoglycan, inhibits glycopeptide synthetase
(2) Ristocetin	mechanism similar to vancomycin			

Table 2.2. *Penicillin-binding proteins of* E. coli *K12*

Penicillin-binding protein	Molecular weight (Mdal)	^{14}C-penicillin G binding (% of total)	Estimated molecules per cell	Genetic locus (min)	Function	Probable effect of inhibition	Relative affinity for				
							Pen G[1]	Mec	Cex	Cep	Cef
1A	95	6 (with 1B)	230	*pon*A(73.5)	Peptidoglycan crosslinking, transpeptidase, likely a 'detour' enzyme with major function when 1B is defective	Probably not essential for growth in presence of 'normal' 1B	+	0	+	++	++
1B	90		230	*pon*B(3.3) (*mrc*(3.3))	Peptidoglycan crosslinking, transpeptidase, peripheral cell wall extension	Inhibition of cell elongation, lysis	+	0	+	+	+
2	66	0.7	20	*rod*A(14.4)	Maintenance of rod cell shape	Large ovoid, osmotically stable cells	+	+	0	0	0
3	60	1.8	50	*fts*I(1.8)	Cell septum formation	Inhibition of septum formation producing filaments	+	0	++	+	+

4	44	4	110	*dacB*(68)	D,D-carboxypeptidase 1B, 1C (1B-membranous, 1C-soluble) low transpeptidase, high endopeptidase activity	Not essential for growth	+	0	+
5	42	65	1800	*dacA*(13.7)	Membrane D,D-carboxy peptidase 1A, relatively insensitive to inhibition by pen G	Not essential for growth. Short half lives for penicillin binding (5–5 min; 6–20 min).	++	0	0
6	40	20	570				++	0	0

[1]Abbreviations: Pen G, penicillin G; Mec, mecillinam; Cex, cephalexin; Cep, cephaloridine; Cef, cefoxitin.

been reported to have four PBPs, *Bacillus subtilis* 5, and *Streptococcus pneumoniae* 4 major and 1 minor. These PBPs are almost surely the initial targets of β-lactams in these and other bacteria. However, as noted subsequently, the interaction of PBPs and β-lactams seems only a partial explanation of the total mechanism of action of β-lactams. Conclusions obtained with studies of *E. coli* PBPs can not be readily transferred to gram-positive organisms as the relationship of PBPs in these two groups of bacteria is unclear.

The most frequent sequence of effects of β-lactams on bacteria is inhibition of cell division, loss of cell viability and cell lysis. Although this sequence is common, it is not absolute as noted for cephaloridine. This drug produces lysis and loss of cell viability at concentrations as low as those producing inhibition of cell division. *Streptococcus sanguis* provides an example of a bacterial species where loss of viability and lysis may not occur or only occur at exceedingly high concentrations of penicillins (these are 'naturally tolerant' to penicillins). However, for most β-lactams and most bacteria the above sequence does take place.

Fig. 2.1. Morphological effects of β-lactam antibiotics on *Escherichia coli* K12

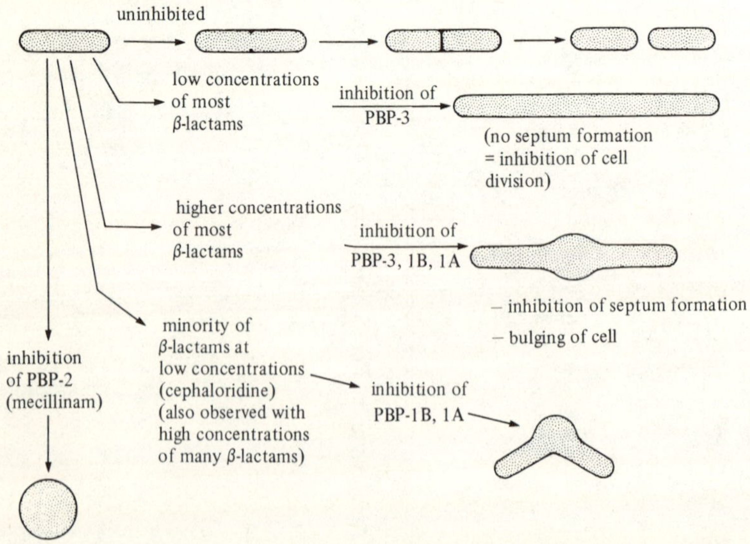

It is probable that inhibition of cell division is the primary action of most β-lactams in most bacteria. This effect is likely achieved by the interaction of the β-lactam ring with the active site of penicillin-sensitive enzymes to form a penicilloyl (acylated) enzyme. This reaction has been proposed to occur because penicillins or cephalosporins (which are both formed from L-cysteinyl-D-valine) are analogs of the donor acyl-D-alanine-D-alanine substrate of these enzymes. The β-lactam bond has been proposed to be a structural analog of the D-alanyl-D-alanine peptide bond (more correctly an analog of a transition state intermediate in the enzymatic cleavage of this bond).

The proposed sequence of reactions is:

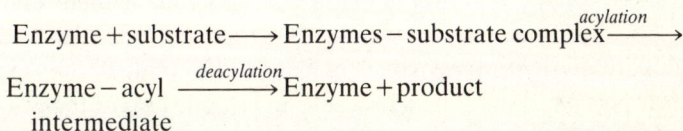

$$\text{Enzyme} + \text{substrate} \longrightarrow \text{Enzymes} - \text{substrate complex} \xrightarrow{\textit{acylation}}$$

$$\underset{\text{intermediate}}{\text{Enzyme} - \text{acyl}} \xrightarrow{\textit{deacylation}} \text{Enzyme} + \text{product}$$

In the case of β-lactams the enzyme–acyl intermediate is only slowly broken down to reactivate the enzyme resulting in inhibition of its action. Products of such reactions include benzyl penicilloic acid and phenylacetylglycine with a thiazoline fragment. The former product is the result of a β-lactamase action reactivating the enzyme.

The substrate analog hypothesis and the formation of an acylated enzyme have not been absolutely proven. However a large body of data is consistent with these proposals to date.

The effect of inhibition of the major penicillin-sensitive enzymes (e.g. PBPs 1 and 3 in *E. coli*) is to cause poorly cross-linked peptidoglycan (due to inhibition of transpeptidation) and/or to prevent incorporation of precursors of the peptidoglycan. These effects are adequate to prevent cell growth. However, the explanation of the irreversible effects of penicillin causing cell death and lysis has been reconsidered recently.

Investigations involving several bacteria but principally *S. pneumoniae* have shown that mutation or certain physiological conditions suppress the activity of the major peptidoglycan hydrolase (an L-alanyl-*N*-acetyl-muramic acid amidase in pneumococci) preventing lysis and reducing loss of viability. These organisms have been described as showing 'antibiotic tolerance'. Other bacteria may possess muramidases and different amidases which act as autolytic enzymes. These enzymes (autolysins) are inhibited by various conditions (choline deprivation and growth in ethanol for pneumococci; lipoteichoic acids of *S. fecalis, B. subtilis, Lactobacillus acidophilus,* inhibit all autolysis of these organisms; low pH for *B. subtilis*) which generally render organisms tolerant to the action of penicillins.

Autolysins may be controlled, at least in many gram-positive bacteria, by complexing with lipoteichoic acids or similar compounds which inhibit the autolysin. The initial action of penicillins on peptidoglycan synthesis may labilize this complex causing release of the inhibitor into the extracellular medium and thus activation of autolytic activity. The triggered enzymatic activity apparently hydrolyzes covalent bonds of peptidoglycan, exposes cytoplasmic membrane and results in osmotic lysis of the cell. This mechanism accounts for lysis. However, even under conditions where autolytic activity is inhibited slow cell death of pneumococci, for example, still occurs. It is possible that the irreversible interaction of β-lactams with enough of essential PBP targets in a single cell could lead to cell death.

Another model of the irreversible effects of penicillin is that interaction of the β-lactams with critical PBPs leads to the lack of balance between cell growth and peptidoglycan traspeptidation (or synthesis). Due to continued growth of cytoplasmic mass, lysis would occur at weakened or deficient sites in the peptidoglycan net. It has been long recognized that growth of the cell is needed for loss of viability and lysis. There is considerable evidence which indicates, however, that the actual requirement is for protein synthesis.

It is difficult to state whether those bacteria that die and lyze with β-lactams do so by identical set of mechanisms. It is likely that the mechanism of irreversibility due to autolysis activity is widespread but may not account for lysis and death of all types of bacteria. The phenomenon of tolerance to penicillins is currently explained by this mechanism. Tolerance has been described for several types of bacteria including *S. pneumoniae, S. mutans, S. sanguis,* clinical isolates of *S. aureus,* group B streptococci and

E. coli. This suggests that the autolytic basis of irreversible penicillin action is widespread.

The mechanism by which mecillinam produces lethality in *E. coli* is unknown and does not seem to be explained by the direct enzymatic autolysin model.

Other inhibitors of the cell wall

The targets, target functions, and apparent mode of inhibition of target functions of other agents are given in Table 2.1. It is probable that inhibitors of the early part of peptidoglycan synthesis such as phosphonomycin and D-cycloserine may also disturb autolysin:inhibitor complexes in the cell. Thus these agents may cause cell death and lysis by the direct enzymatic destruction of peptidoglycan similar to that proposed to occur for β-lactams.

Combination of fluoro-D-alanine and either D-cycloserine or the agent pentizadone (inhibition of D-alanine: D-alanine synthetase) may effectively inhibit bacteria. The combination has a wide spectrum of antibacterial activity and provides a double blockade. The latter prevents the self reversal of action seen with the competitive inhibition due to fluoro-D-alanine alone.

Bacitracin probably binds in association with a divalent cation to the pyrophosphate moiety of the C_{55}-isoprenyl pyrophosphate carrier involved in transfer of UDP-acetylmuramyl-pentapeptide precursor to peptidoglycan. This complex prevents dephosphorylation of the carrier to the monophosphate form needed to bind the peptidoglycan precursor. Thus peptidoglycan synthesis is inhibited. Much experimental evidence supports this interaction. Bacitracin also acts to alter permeability of protoplasts which do not require peptidoglycan synthesis. Thus the effect of bacitracin may be to prevent peptidoglycan synthesis or cause membrane leakage or both. Unquestionably the drug possesses a high affinity for C_{55}-isoprenyl pyrophosphate. It is likely that the drug–lipid association is the basis of both peptidoglycan and permeability effects.

Phosphonopeptides have recently been examined as antibacterial agents. One of the most active of these alaphosphin, is shown in Fig. 2.2.

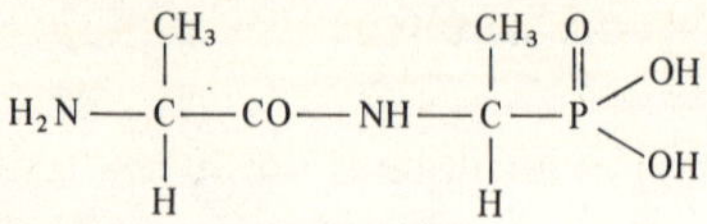

Fig. 2.2. Alaphosphin.

This compound is transported into cells by stereospecific peptide permeases and hydrolyzed by aminopeptidases to L-alanylphosphonic acid. The latter compound is concentrated 100 to 1000 fold and inhibits cell-wall synthesis by competitive inhibition of alanine racemase. It is also secondarily inhibits the UDP-*N*-acetylmuramyl-L-alanine synthetase.

Alaphosphin is frequently synergistic with other inhibitors of cell wall synthesis.

Cytoplasmic membranes as targets

A variety of antibacterial agents act to modify the function of some component of the cytoplasmic membrane. Many of these agents have poor selective toxicity and cannot be used for systemic infections. Other agents can be used for systemic infections but only with a significant risk of toxic effects on specific host tissues. The major agents available for systemic use as antibacterial agents are polymyxins.

Agents producing membrane disorganization

Polymyxin B, and Polymyxin E_1 (Colistin A) structures are shown in the appendix. These agents bind electrostatically in place of divalent cations to negatively charged amphipathic molecules like cardiolipin in the membrane. It is probable that the hydrophobic tail of polymyxins enters the hydrophobic domain of the membrane and interacts through hydrophobic interactions with membrane fatty acids. The effects of the above interactions likely disrupt membrane lipid packing, increasing the permeability of the membrane. These changes are shown diagrammatically in Fig. 2.3. The reason for antagonism of polymyxins by divalent cations is also shown in that Fig. Both polymyxins B and E are neurotoxic and nephrotoxic. These side effects limit the systemic doses of the drugs that can be used. The action of polymyxins on the outer membrane is discussed in Chapter 5.

Several other agents also act on membranes resulting in some degree of membrane disorganization. These agents contain separate hydrophilic and hydrophobic portions. Their effects seem to be secondary to their partition between a charged environment represented by the heads of phospholipids and a hydrophobic environment represented by the packed fatty acids of the membrane. These agents include many cationic (e.g. acetyl trimethyl ammonium bromide) and anionic (sodium dodecyl sulfate, some phenols) surface active agents. Some polyenes such as filipin may act to cause disorganization of the membrane rather

Fig. 2.3. Model for the action of polymyxins on bacterial membranes. The polymyxin molecule consists of a heptapeptide positively charged head and a lipophilic tail. The head binds to negatively charged polar components of phospholipids like cardiolipin. Normally divalent cations like Mg^{2+} and Ca^{2+} act as bridges between adjacent phospholipids. Insertion of the polar heptapeptide displaces the cations and disturbs normal packing of membrane phospholipids. The fatty acid tail enters the hydrophobic portion of the membrane. These effects likely combine to significantly modify packing of phospholipids within the membrane leading to increased permeability for hydrophobic materials. Divalent cations are effective antagonists of polymyxin because they compete with polymyxins for negatively charged phospholipid binding sites. Reproduced, with permission, from the *Annual Review of Biochemistry*, vol. 46. © 1977 by Annual Reviews Inc.

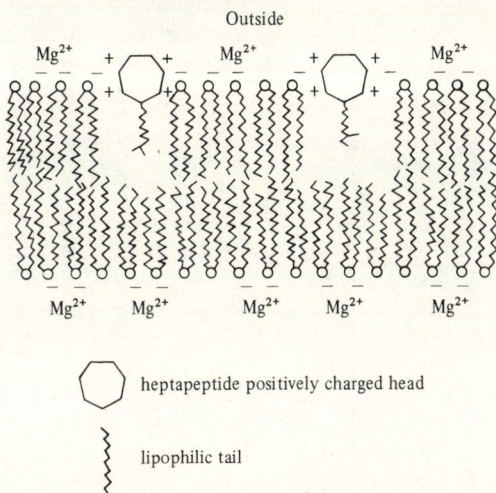

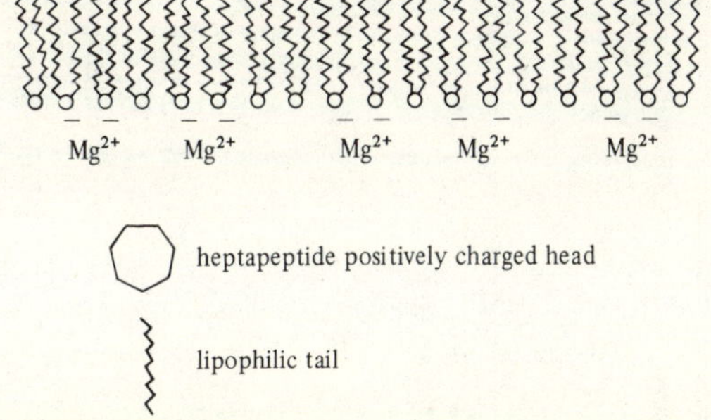

Table 2.3. *Ionophores*

Class	Compound	Origin	Antibiotic activity		Structure	Ion selectivity	Comments
			Gram-positive	Fungal			
Channel formation	Polyenes	*Streptomyces* sp.	–	+(R)[a]	polyenes	Very little, somewhat anionic specific	Amphotericin B – used as systemic antifungal agent Nystatin – topical antifungal agent Others used uncommonly
	Gramicidins A,B,C.	*Bacillus brevis*	+		peptides gramacidin A	$H > Rb > NH_4 > K > Na > Li$; K/Na = 1.8[b]	
	Alamethicin	*Trichoderma viride*	+/–	–(most)		$K > Rb > Na > Li$; K/Na = 1.7; Ca^{2+}	Alamethicin and monoazomycin depend on the membrane potential (voltage) to form channels
	Monoazomycin	*Streptomyces* sp.	+/–	–	unknown	Na/K = ~ 1; Ca^{2+}	

Neutral carriers	Valinomycin	*Streptomyces fulvissimus*	+	+(R)[a]	cyclodepsipeptide	$Rb \gg K \gg NH_4 > Na > Li$; K/Na = 10 000	Not used clinically Specific K^+ ionophore
	Nactins	*Streptomyces* sp.	+		macrotetralide	$NH_4 > K > Rb > Na > Li$; K/Na = 210	
	Enniatins	*Fusarium orthoceras*	+	+	cyclodepsipeptide	A–$K > Na$; $Rb, Cs > Li$; B–$NH_4 > K > Rb > Na > Li$; K/Na = 37	
	Diamide ligands					$Ca > Sr > Ba > Mg$; K/Na = 0.2	
Carboxylic carriers	Nigericin	*Streptomyces* sp.	+	+	polyether	H; $K > Rb > Na > Cs > Li$; K/Na = 45	Structure of nigericin, and monensin is similar
	Monensin	*Streptomyces cinnamonensis*	+		polyether	H; $Na > K > Rb > Li > Cs$; K/Na = 0.1	
	Dianemycin	*Streptomyces hydroscopius*	+/–		polyether monoglycoside	H; $Na > K > Rb, Cs > Li$; K/Na = 0.5	
	Lasalocid	*Streptomyces* sp.	+		polyether	H; $Cs > Rb, K > Na > Li$; $Ba > Ca > Mn > Sr > Mg$	Used for coccidia infections in chickens
	Calcimycin	*Streptomyces chartreusensis*	+		polyether	H; $Li \gg Na > K$; K/Na = 0.1; $Mn \gg Ca > Mg > Sr > Ba$	Calcium ionophore, two molecules per Ca^{2+}
Proton conductor	Dinitrophenol,	phenylhydrazones, salicylanilides	+		–	H	Powerful uncouplers
	Phenols,	benzimidazole	+	+	–	H	Weak uncouplers

[a]R–some fungi are resistant.
[b]Numbers provided indicate the ratio of selectivity for Na^+ relative to K^+.

than to cause channel formation as amphotericin B and nystatin do.

Agents acting as ionophores

The problem with almost all ionophores is that they have little selectivity for bacterial over mammalian cell membranes. Thus, they tend to be highly toxic and are limited in use. The outer membranes of many gram-negative bacteria probably exclude most ionophores suggesting these molecules permeate this complex through the pores formed from major outer membrane proteins (see Chapter 5). This mechanism accounts for the resistance of many gram-negative bacteria to most ionophores.

In general 'true' ionophores act to replace the water mantle of an ion and transmit it across the membrane as a hydrophobic ion–ionophore complex. A second mechanism is the formation of a transmembranous channel by various agents which are not strictly ionophores but are often described as such. (See Table 2.3.)

Channel forming ionophores. Polyene antibiotics have been widely used topically and systemically to treat fungal infections. Amphotericin B, in particular, has been used systemically. A therapeutic course of the drug is usually followed by a permanent reduction in glomerular filtration rate. Acute toxic reactions also occur and require the use of small initial test doses.

Polyenes are active only on membranes which contain sterols. Bacteria are insensitive because they lack sterols. Mycoplasma membranes may contain sterols or may, like *Acholeplasma laidlawii* incorporate sterols when grown in their presence and thus be inhibited by polyenes. Amphotericin B has greater affinity for ergosterol (found in fungi) than cholesterol.

Amphotericin B associates with molecules of cholesterol (or other sterol) in the membrane to form a transmembranous channel of a diameter about 4 Å. It has been proposed that eight molecules each of amphotericin and cholesterol together form a half pore. A whole pore results when two halves align correctly. The little specificity that these pores have for ions is for anions. Glucose also traverses the pores.

Amphotericin B methyl ester is a derivative with greater water solubility but with little reduction of *in vitro* activity. However, experimental infections have indicated decreased *in vivo* activity for unknown reasons. Hearing loss has occurred in humans. Its role in treatment of human fungal infections is still under assessment.

Gramicidins A, B, and C are polar peptides with blocked terminal amino and carboxyl groups and, thus, have no charge. Gramicidins S and J are not channel formers but detergents.

The channel forming gramicidins form a channel from a gramicidin dimer and probably form a narrow 4 Å channel with preferential passage of monovalent cations.

Other channel forming ionophores are noted in Table 2.3. These include the voltage-dependent channel formers. These are interesting agents because the so called 'gated channels' strongly resemble those of cation channels in axons and other excitable membranes.

Filipin a polyene induces non-specific permeability changes and cell lysis. Thus it does not seem to act as a channel former.

True ionophores. (*a*) *Neutral carriers.* These represent a group of structurally different compounds which are neutral before interacting with a cation. After interaction they become a charged cation–ionophore complex and move across the membrane in response to their transmembranous electrical potential. The carrier, freed of the cation, diffuses back to the opposite side of the membrane. This process tends to dissipate the electrical potential of the cell.

As shown in Table 2.3 these compounds exhibit selectivity for different cations. None of them are used in clinical medicine as antibiotics. They have been powerful tools particularly in the study of the chemiosmotic basis of ATP synthesis and transport of compounds.

(*b*) *Carboxylic carriers.* These compounds possess a carboxylic acid group. They move in one direction as a neutral cation–anion complex and in the reverse direction as an undissociated acid. Thus a cation–proton exchange occurs and results in an electrically silent cation transmission.

Some of these compounds complex and transport divalent cations. Lasalocid (X-537A) also complexes biogenic amines. Unlike monensin, nigericin and dianemycin this agent is too small to form a cavity for a dehydrated cation. Cations or biogenic amines bind to a polar face of a disk-type structure. This agent has been used for coccidial infections of birds. Calcimycin (A-23187) sandwiches calcium between two equivalent anionic molecules. It is the best Ca^{2+} ionophore available at the time of this writing in spite of relatively poor selectivity over Mg^{2+}. The interest of calcium ionophores lies in the fact that most cells extrude calcium. Small change in cytosol concentration of Ca^{2+} modify the activity of many biological functions as Ca^{2+} regulates many intracellular processes.

(*c*) *Other carriers.* A series of compounds act as proton conductors all of which are toxic to human cells. These include 2,4-dinitrophenol, phenylhydrazones and salicylanilides. The agents collapse the transmembranous proton gradient and act to uncouple oxidative phosphorylation.

Agents inhibiting the action of bacterial membrane ATPase

Chlorhexidine inhibits bacterial ATPase. K^+ loss from cells is a very early manifestation of the action of biguanides like chlorhexidine. When the ATPase is inhibited the intracellular concentration of K^+ can not be maintained. It is also possible that K^+ leakage may occur through a direct effect of chlorhexidine on the cytoplasmic membrane. A variety of other agents also inhibit the ATPase but are generally highly toxic and often have a limited antibacterial spectrum.

Inhibitors of nucleic acid synthesis

A large number of agents with antimicrobial and/or antitumor properties belong to this group. Their mechanisms of action are given in Table 2.4.

Rifampicin is a semi-synthetic derivative of a rifamycin. The modification (Table 2.4) resulted in a marked increase in antibacterial activity as an inhibitor of RNA polymerase. The drug has been used against gram-positive bacteria and mycobacteria and in the treatment of meningococcal carriers. It obtains high

Table 2.4. *Inhibitors of nucleic acid synthesis*

Agents	Primary mechanism of action
1. Agents interfering with nucleotide metabolism	
(*a*) Nucleotide synthesis	
trimethoprim	Very high affinity for bacterial dihydrofolate reductase. Potent selective competitive inhibitor of this enzyme.
pyrimethamine	High affinity for malarial dihydrofolate reductase. Potent selective competitive inhibition of this enzyme. Also inhibits *Toxoplasma gondii*.
azaserine	Structural analog of glutamine. Inhibits, particularly, the phosphoribosyl-formyl-glycineamidine synthetase step in purine synthesis.
other antitumor agents (6-azauridine, xylosyladenine and others)	Purine and pyrimidine synthesis.
(*b*) Nucleotide interconversion	
hadacin	Analog of aspartic acid, inhibits adenylo-succinate synthetase (synthesis of adenine nucleotides).
6-mercaptopurine psicofuranine	Converted to nucleotide analog acts as a nucleoside and inhibits XMP aminase non-competitively.
5-fluorouracil 5-fluorodeoxyuridine	Converted to 5-fluorodeoxy-UMP, inhibit thymidylate synthetase.
(*c*) Inhibit nucleotide incorporation	
cystosine arabinoside (ara-C), adenine arabinoside (ara-A)	The mechanism of action is not finally established. DNA polymerases, reverse transcriptases and other enzymes are inhibited. In one model ara-ATP and ara-CTP (triphosphate derivatives) competitively inhibit deoxynucleoside triphosphosphate incorporation into DNA. In another model the analogs are incorporated into DNA and impede further DNA synthesis by several possible mechanisms.

Table 2.4. *(Contd.)*

Agents	Primary mechanism of action
(*d*) Incorporation of analogs into DNA or RNA	
5-bromodeoxyuridine 5-iododeoxyuridine	Probably incorporated into DNA in place of thymidine.
adenine arabinoside (ara A)	Deaminated to arabinosyl hypoxanthine, enters cells and is phosphorylated. It is a competitive inhibitor of DNA polymerase (especially viral) and is incorporated into DNA.
8-azaguanine	Analog of guanine; incorporated into RNA.
cordycepin (3′-deoxyadenosine)	Converted to triphosphate and incorporated into 3′ end of RNA preventing further elongation.
5-fluorocytosine	Converted to triphosphate and incorporated into RNA which may have impaired function.
2. Agents inhibiting template function of DNA	
(*a*) intercalating agents	
acridines, ethidium bromide	Insertion of flat aromatic molecules between adjacent base-pairs of the DNA double helix with consequent local un-winding of the helix. Closed circular DNA (CCC) has a higher affinity for these agents than linear DNA. At saturating levels, the right-handed supercoil unwinds and assumes a reversed supercoil with reduced affinity for intercalating agents. This results in less drug bound by CCC DNA and a smaller reduction in density as compared to linear and open circular DNA. The higher affinity of right-handed CCC helices at limiting concentrations may account for the preferential loss of extra-chromosomal DNA (curing) due to these drugs. Intercalating agents also cause frameshift mutations.
chloroquine	Intercalation with DNA and subsequent inhibition of DNA synthesis.
miracil D many others	

Table 2.4. *(Contd.)*

Agents	Primary mechanism of action
(*b*) actinomycin D	Binding to DNA (guanine is important in the binding site) and inhibition of DNA-dependent RNA polymerase synthesis of RNA.
(*c*) chromomycin mithramycin olivomycin	Binding to G-C rich regions of DNA to form a non-covalent complex and inhibition of particularly RNA synthesis in mammalian cells and both RNA and DNA synthesis in certain bacterial cells.
(*d*) anthracyclines (examples daunomycin, nogalamycin, cinerubins and others)	Binding to DNA to form a non-covalent complex (intercalate), inhibit RNA and DNA synthesis.
(*e*) covalent complex formation mitomycin C other drugs (e.g. antitumor alkylating agents)	Mitomycin C is activated by reduction to form a bifunctional alkylating agent. It forms covalent bonds with DNA and some molecules cross-link DNA strands. Selective inhibition of DNA synthesis and in some cells DNA degradation.
3. Agents inhibiting polymerases and other enzymes in DNA synthesis	
(*a*) Inhibitors of RNA polymerase	
rifamycins streptovaricin	Rifampicin is 3-(4-methylpiperazinyl-iminomethyl) derivative of rifamycin SV. Rifampicin binds to the β subunit of bacterial RNA polymerase. The *E. coli* enzyme is 50% inhibited at 2×10^{-8} M rifampicin. Causes inhibition of initiation of RNA synthesis. Rifamide is rifamycin B diethylamide and is less active than rifampicin.
streptolydigin	Inhibits bacterial RNA polymerase. Inhibits the chain-polymerization reaction.
(*b*) Inhibitors of DNA replication	
nalidixic acid	Inhibits nal subunit of DNA gyrase, the enzyme inducing negative superhelical turns into CCC DNA. Probably acts as a nicking-closing activity. MW in *E. coli* – 100 000 to 105 000

Table 2.4. *(Contd.)*

Agents	Primary mechanism of action
novobiocin coumermycin A	Competitive inhibitor of ATP binding to cou subunit of DNA gyrase, affinity 10 000 times that of ATP. Probably introduces twists in DNA molecules.
edeine A	Inhibits DNA synthesis probably by inhibiting bacterial DNA polymerases II and III.
phenethyl alcohol	Mechanism unknown.
phosphonoacetate phosphonoformate	Inhibit Herpes virus DNA polymerase as well as other viral DNA polymerases (Vaccinia, cytomegalovirus).

biliary concentrations and may be of use in biliary tract infections.

Nalidixic acid, an inhibitor of DNA-gyrase (Table 2.4), is active principally against gram-negative bacteria. It is used almost entirely for treatment of infections of the urinary tract. Novobiocin is a drug active on the cou subunit of the DNA gyrase enzyme not inhibited by nalidixic acid. It was once widely used as an anti-staphylococcal agent. However, the development of superior agents, the high albumin-binding of novobiocin, its impairment of hepatic function and a high incidence of skin eruptions have caused the drug to be little used.

Chloroquine, quinacrine and primaquin are effective antimalarial agents and active against some other protozoa. Chloroquine is active on the erythrocytic cycle of malarial parasites but not on the exo-erythrocytic cycle. The latter is susceptible to primaquin.

Several antiviral agents (see Table 2.4) have been shown to be effective for topical therapy of herpes keratoconjunctivitis. These include 5-iododeoxyuridine (IDU), adenine arabinoside (ara-A) and trifluorothymidine. However, treatment of mucocutaneous disease due to Herpes simplex has been much less effective. The use of IDU in dimethylsulfoxide (DMSO) has been promising. This suggests that one of the problems of treatment may be difficulty of IDU in entering cells. Decreased viral shedding from mucocutaneous herpes has been obtained with ara-A although not with genital herpes.

Ara-A has shown therapeutic efficacy for herpes encephalitis. In addition this drug has produced acceleration of cutaneous healing of lesions due to herpes zoster in immunocompromised patients. It is likely that hypoxanthine arabinoside formed by deamination of ara-A is the active drug. Monophosphorylated ara-A has been recently produced to improve the solubility of this compound and, thus, reduce the volume of fluid needed for administration of ara-A.

Several agents are under current evaluation for inhibition and potential therapy of Herpes simplex 1 and 2 viruses. Some of the most promising of these are acycloguanosine, 2′-fluoro-5-iodoaracytosine, E-5-(2′bromovinyl)-2-deoxyuridine, E-5-(2-iodovinyl)-2′-deoxyuridine and 5-vinyl-2′-deoxyuridine. These agents show much reduced activity in viral mutants deficient in a viral specified thymidine kinase. Thus it is likely they must be phosphorylated and act as described for ara-C and ara-A in Table 2.4, part 1 (*c*). The agent 2-deoxy-D-glucose is active on Herpes simplex 2 but likely acts to interfere with synthesis of a major glycosylated polypeptide of the virus.

5-fluorocytosine is used orally for the treatment of systemic fungal infections. It is widely distributed in body tissues and is relatively non-toxic.

Protein synthesis as a target

Bacterial protein synthesis utilizes 70 S ribosomes which differ from the 80 S eukaryotic non-mitochondrial ribosomes. Mitochondrial protein synthesis, however, involves prokaryotic-type ribosomes. Many antibacterial agents exert selective toxicity through their action on 70 S ribosomes, subunits thereof or associated factors.

The effects described for each of the inhibitors are primarily effects on protein synthesis. In most cases additional effects have also been described at higher drug concentrations. In some cases the basis of action has not been finalized and the views given are those consistent with most authors' reports.

Inhibitors involving the larger ribosomal subunit

Chloramphenicol. The structure of chloramphenicol is given in the appendix. Only the D-threo isomer has strong antibacterial activity. The aromatic ring system and the acyl side chain can undergo a variety of substitutions without loss of activity. Substitutions of the propanediol component are, however, very restricted without loss of activity.

Chloramphenicol inhibits protein synthesis resulting in bacteriostasis at usual concentrations. Ribosomal binding is reversible as the drug can be removed from both ribosomes and whole cells by washing. This seems to be the major explanation as to why this agent does not cause irreversible inhibition of protein synthesis and cell growth Not all protein synthesis is equally sensitive to inhibition by chloramphenicol. Some proteins involved in DNA and phage replication are synthesized in its presence. Resistance of some protein synthesis seems due to the base composition of mRNA or ribosomes associated with the cell membrane that may bind less chloramphenicol. Mitochondrial and chloroplast protein synthesis is inhibited by chloramphenicol.

Chloramphenicol has been shown by *in vivo* and *in vitro* affinity labelling with monoiodoamphenicol and monobromoamphenicol to bind to several ribosomal proteins. The primary target seems protein L 16 (L = large ribosomal subunit) which apparently forms part of the A-site (aminoacyl-tRNA site) of the peptidyl transferase center. This likely represents the major binding protein of the higher affinity of two binding sites. Secondary binding representing a lower affinity site may involve proteins (L 24, L 27, ?L 2) that are part of the P-site (peptidyl-tRNA site) of the peptidyl transferase center. Chloramphenicol inhibits binding of aminoacyl-tRNA and peptide bond formation. A protein (S 6) of the small ribosomal protein is also labelled by *in vivo* affinity labelling but its role remains unclear.

Several other antibiotics inhibit chloramphenicol ribosomal binding including puromycin, many macrolides including erythromycin, lincomycin and several others. Chloramphenicol, however, does not inhibit erythromycin binding and does not occupy the identical site of puromycin.

Chloramphenicol penetrates very well into a wide variety of

tissues including the brain. Much of this excellent tissue entry is related to the high lipid solubility of the drug. It is active on a wide spectrum of bacteria including many anaerobes. Currently it is a drug of choice for typhoid fever and meningitis due to ampicillin-resistant *H. influenzae.* Unfortunately the drug suppresses formation of cells of the bone marrow particularly erythrocytes on a dose-related basis. In infants detoxification by formation of the glucuronide is impaired and toxicity as the gray-syndrome is more probable. An irreversible fatal suppression of bone marrow occurs in 1 in 10 000 to 1 in 600 000 cases. The last toxicity is thought to be less common with parenteral and conjunctival routes of administration.

Macrolides and lincomycins. These groups of agents inhibit protein synthesis by binding to the 50 S ribosomal subunit. Lincomycin and erythromycin apparently bind to different sites which are functionally related.

Lincomycin binding requires K^+ or NH_4^+ and Ca^{2+} or Mg^{2+}. It binds probably to a protein or proteins involved in the ribosomal A site and primarily interferes with aminoacyl-tRNA binding to that site. Protein L 16 is probably involved. Some mutants of *E. coli* resistant to lincomycin have defective L 6 proteins. Thus, peptide synthesis is prevented. Lincomycin inhibits the 'puromycin reaction' and the 'fragment' reaction (involving reaction between terminal fragments of fMet-$tRNA_f$ and puromycin) which are measures of peptidyl transferase activity.

Lincomycin and clindamycin are active on most gram-positive bacteria but are not very active on many gram-negative bacteria. However, they do inhibit many gram-negative anaerobes and *H. influenzae* and species of *Neisseria.* The lack of activity on *E. coli* may be due to a failure to form stable complexes with ribosomes except at high drug to ribosome ratios.

Ribosomal binding of lincomycin may involve principally hydrophobic interactions. Lincomycin binds to ribosomes less strongly than does erythromycin. As a result erythromycin can antagonize the action of lincomycin by displacing it from its binding site even though the binding sites are not identical.

Clindamycin is 7-chloro-7-deoxylincomycin. It has many properties in common with the parent drug but is generally more active than lincomycin. It is very active on many anaerobic bacteria and is better absorbed as the oral hydrochloride. The parenteral preparation of clindamycin is clindamycin phosphate. The use of clindamycin has occasionally been associated with pseudomembranous enterocolitis resulting from a toxin of the organism *Clostridium difficile.* This clinical entity also results from the use of other antibiotics particularly ampicillin.

Erythromycins are antibiotics containing a macrocyclic lactone ring with a neutral and an amino-deoxy sugar in glycosidic linkage. At least one sugar is required for antibacterial activity. The common erythromycin in wide use and referred to as erythromycin is erythromycin A (see appendix). Maximal growth inhibition occurs at about pH 8.5. The pKa of erythromycin is 8.6 suggesting the most active form is the non-protonated species. Lipophilicity is maximal at pH 8.6; maximal activity may depend on lipid solubility in the bacterial cytoplasmic membrane.

Erythromycin is active on many gram-positive bacteria including some atypical mycobacteria. It is also active on *Mycoplasma pneumoniae, Ureaplasma urealyticum, Legionella pneumophila* and Chlamydiae although not necessarily the agent of first choice for these organisms. These susceptibilities have extended the use of erythromycin for 'primary atypical pneumonia' and non-gonococcal urethritis. *Neisseria* and *Hemophilus* species may be susceptible as are many anaerobic bacteria. It should be noted that middle ear concentrations will usually be inadequate to inhibit *H. influenzae.* Most gram-negative enteric bacteria are relatively resistant.

Erythromycin binds to ribosomes as one molecule per 50S ribosomal subunit at saturating concentrations although additional less firm binding sites also exist. Mutants resistant to erythromycin show variable cross resistance to other macrolides and even lincomycin. Ribosomal binding sites are not identical for these compounds but are closely associated. In several mutants a single protein from the large ribosomal subunit has been shown to have an altered amino acid sequence in a limited portion of that protein.

Erythromycin inhibits protein synthesis and causes bacteriostasis although it may be bactericidal at higher concentrations. The primary effect of the drug is probably to interfere with positioning of peptidyl-tRNA, particularly those with large peptides and to inhibit translocation. Other effects are seen with higher drug concentrations. Monovalent cations (NH_4^+ or K^+) are absolutely required for its action. It is likely that erythromycin must bind to the 50 S subunit before it enters a polyribosomal complex. Eukaryotic non-mitochondrial protein synthesis is not inhibited but mitochondrial protein synthesis can be.

Toxicity to erythromycin is extremely unusual. Reversible cholestatic jaundice is an unusual manifestation of the use of erythromycin estolate but not other preparations. Esters of erythromycin base are used to reduce acid lability (relative to the base), for improved taste, and to prepare stable solutions in water.

Other clinically important macrolides include oleandomycin, spiramycin and rosaramicin. In general oleandomycin is comparable to erythromycin. Spiramycin is useful as a substitute for pyrimethamine for toxoplasmosis in pregnancy. Rosaramicin is a new macrolide produced by *Micromonospora rosaria.* It is probably more active than erythromycin against *H. influenzae,* some anaerobes, *N. gonorrhoeae, C. trachomatis* and many Enterobacteriaceae although not strains of *Proteus.* Like erythromycin, activity declines with acid pH. It has been shown less efficacious than penicillin G in experimental pneumococcal meningitis. Its role in therapy remains to be determined.

Other inhibitors involving the 50 S ribosome. Puromycin acts as an analog of the 3′terminal aminoacyl adenosine end of aminoacyl-tRNA and binds to the ribosomal A site. Peptide formation occurs between it and the peptidyl-tRNA in the P site. The peptidyl puromycin produced is released from the ribosome due to a lack in puromycin of those structures needed to bind firmly to the ribosome. Thus protein synthesis is inhibited. The agent is not used for humans because of toxicity.

Other inhibitors such as gougerotin, sparsomycin, anisomycin, streptogramin A, thiostrepton and other related agents that act on the 50 S ribosome are not used in clinical medicine for treatment of bacterial infections.

Emetine has been used as an antiparasitic agent principally

against *Entamoeba histolytica.* Its primary effect is to inhibit movement of mRNA along the ribosome and cause formation of 80 S emetine complexes. It, thus, produces inhibition of protein synthesis in eukaryotes. Cyclohexidine also inhibits eukaryotic protein synthesis by blocking initiation perhaps by inhibition of binding of initiating tRNA to ribosomes or the junction of 60 S subunit to initiation complexes. Chain elongation is also blocked at somewhat higher concentrations.

Inhibitors of the smaller ribosomal subunit

Tetracyclines. Tetracyclines are a family of antibiotics of which the structure of the most commonly used members are given in the appendix. These compounds are generally used as hydrochlorides to increase water solubility. Minocycline is the most potent of the group and is also the most lipophilic. Its enhanced activity may be due to its solubility in membranes and thus improved penetration to the cellular target.

Tetracyclines have a wide spectrum of antibacterial activity including most bacteria, *Mycoplasma, Ureaplasma, Rickettsia, Chlamydia* and *Coxiella.* It remains a primary agent for gonorrhoea, brucellosis, granuloma inguinale, melioidosis, cholera, many chlamydial infections, *M. pneumoniae,* typhus, Q fever, acne and a secondary agent for numerous other infections.

Resistance to tetracyclines has developed in many bacteria including *S. pneumoniae, S. pyogenes* and the *Bacteroides fragilis* group of bacteria thus limiting the use of these agents in several clinical conditions.

Toxicity involves tetracycline deposition in developing bone and the inhibition of fetal osteogenesis. This effect is probably the result of inhibition of calcification although inhibition of protein synthesis has not been absolutely excluded. Tetracycline does penetrate the placenta and thus fetal bone development can be slowed. Tetracyclines also stain teeth in utero after the fifth month of pregnancy and may stain anterior-teeth up to three years of age. These agents accumulate during renal failure with the exception of doxycycline (in most cases). If accumulated they exert an anti-anabolic action and have a deleterious effect on the course of the patient. They may also cause liver damage and have been associated with superinfection of the gastro-intestinal tract.

The primary mechanism of action of tetracyclines is to prevent enzymatic and non-enzymatic binding of aminoacyl-tRNA to the A site of the ribosome. Although binding to the P site can be prevented, the major effect is at the A site. Tetracycline chelates divalent cations including Mg^{2+}. The 1:1 chelate of tetracycline and Mg^{2+} binds to ribosomes. It seems that one tetracycline molecule binds strongly to the 70S ribosome and is responsible for the inhibition of aminoacyl-tRNA binding. Weaker binding occurs to the 30S subunit which does prevent aminoacyl-tRNA binding. Weak binding to the 70S ribosome is not inhibitory. Tetracycline also inhibits polypeptide chain termination by inhibition of the interaction of termination factors RF_1 or RF_2 with the termination codons. At low concentrations tetracyclines stimulate RNA synthesis by inhibiting the synthesis of guanosine tetra- and penta-phosphates which are responsible for the 'stringent' control of RNA synthesis by amino acids.

Aminoglycosides-streptomycin and dihydrostreptomycin. Although a good deal is known about the interaction of streptomycin with the ribosome and subsequent effects on protein synthesis, the mechanism by which bacterial death results from streptomycin or other aminoglycosides remains unknown. Streptomycin interacts with the 30S ribosomal subunit. The interaction involves the S12 ribosomal protein but the drug does not bind directly to that protein. However the S12 protein does control whether binding occurs or not. The exact binding protein or site remains undetermined. Multiple investigations have produced a series of divergent results indicating binding to several proteins of the smaller ribosomal subunit. It has also been suggested that streptomycin and dihydrostreptomycin (DHS) binding involves a direct interaction with 16S RNA although the characteristics of binding for the two drugs were different. Wallace *et al.* have suggested that binding is to the region of the 30S ribosome between the platform and the cleft. They feel this is supported by evidence which connects this region with several sites of function which are known to be affected by streptomycin.

Ribosomes appear to have both high and low affinity binding

sites for dihydrostreptomycin (DHS). The high affinity site involves the binding of one molecule per ribosome up to a concentration of 10^{-5} M and requires magnesium. The dissociation constant at 25 °C for the bound complex is 9.4×10^{-8} M. The effect of the interaction between DHS or streptomycin and the ribosome is to produce a change in ribosomal conformation which results in a series of secondary effects on protein synthesis.

DHS and streptomycin produce complete inhibition of protein synthesis when mixed with ribosomes involved in initiating protein synthesis. However, when mixed with preformed polysomes (i.e. ribosomes engaged in chain elongation) these agents only slow protein synthesis and cause misreading on such ribosomes. The molecular mechanism for misreading remains unknown but misreading does occur with synthetic and natural messenger RNAs. The effect of misreading can be to produce phenotypic suppression at sublethal concentrations of streptomycin.

When cells contain a mixture of streptomycin sensitive and resistant ribosomes, sensitivity to streptomycin is dominant. This is possibly due to the cycle of abortive initiations that occurs in the presence of streptomycin. When cells are exposed to streptomycin, initiation of protein synthesis is inhibited and ribosomes are released from messenger RNA and can take part in re-initiation of protein synthesis. The released 70 S ribosomes have increased stability to IF-3 factor mediated dissociation. These unusually stable 70 S particles accumulate in cells treated with streptomycin and have been repeatedly observed. It is proposed that in the presence of a mixture of streptomycin-susceptible and resistant ribosomes, cyclic abortive initiation includes sensitive ribosomes that occupy initiation sites and, thus, exclude resistant ribosomes. In a mixture of sensitive and resistant ribosomes, sensitive ribosomes would bind streptomycin and take part in cyclic abortive initiation. These initiations would occupy initiation sites for a very large fraction of time relative to the time that initiation sites would be occupied by resistant ribosomes which did not bind streptomycin and did not take part in the cyclic abortive initiations.

The binding of streptomycin and ribosomes and the subsequent change in conformation of the ribosome produces many effects on protein synthesis. These include: an inhibition of dissociation of free ribosomes (30 S and 50 S subunits), a decreased stability of initiation complexes and distortion of the P site, distortion of the A site, inhibition of ribosome release from messenger RNA after peptide release, inhibition of peptide termination mediated by peptide release factor and finally a possible effect on translocation.

Mutations to streptomycin and DHS resistance involving the *str*A locus always show change in the S12 protein and involve 1 of only 2 amino acids in that protein. One streptomycin resistant mutant has been described which involves two changes in the S12 protein.

Lethality resulting from streptomycin remains an enigma. It is known that the earliest effect of streptomycin is to cause initial low-grade potassium leakage with subsequent enhanced potassium leakage occurring about the same time that there is evidence of inhibition of protein synthesis. Leakage of potassium is subsequently followed by leakage of higher molecular weight material from the cell. About the time that the potassium leakage accelerates there is a stimulation of RNA synthesis. Eventually, cell respiration is extensively inhibited. It is possible that the effects on protein synthesis may be adequate to cause cell death. However, it remains a strong possibilty that the effects on membrane permeability in association with the ribosomal interaction are responsible for loss of viability.

The activity of streptomycin and other aminoglycosides is reduced by anaerobic conditions and low pH. The reason for these effects is given in Chapter 5.

Other aminoglycosides and aminocyclitols. Numerous additional aminoglycosides currently are in some degree of clinical use or under evaluation. These include: kanamycin, tobramycin, amikacin, dideoxykanamycin B, 6′-*N*-methyldideoxykanamycin B, UK 31214, gentamicins (C_1, C_{1a}, C_2), sisomicin, netilmicin, neomycin B, paramomycins, lividomycins, ribostamycin, butirosins, kasugamycin and others.

Aminoglycosides have a wide spectrum of antibacterial activity affecting many of the clinically important gram-negative and gram-positive bacteria including mycobacteria. They are inactive on most anaerobes as they are not transported into most such bacteria. In addition they have relatively low activity on streptococci due also to poor transport (see Chapter 5).

Aminoglycosides including streptomycin have their major toxicity on renal, vestibular or hearing functions. The extent of toxicity for each of these varies depending on the agent.

In general aminoglycosides inhibit protein synthesis and cause misreading of messenger RNA. Neomycin, kanamycin and gentamicin produce higher levels of misreading than streptomycin. They show an increase in misreading as drug concentration is raised from 10^{-6} M to 10^{-4} M suggesting an interaction of ribosome and drug at more than one site. Aminoglycoside interaction with the ribosome is altered by mutations affecting mainly 30 S ribosomal proteins (see Table 3.4, Chapter 3). To date gentamicin resistance has been associated with a mutation of protein L6 from the 50 S subunit in combination with membrane energy mutants. The effects of aminoglycosides are influenced by several structural aspects of the drugs. The disaccharide or pseudodisaccharide portions seem essential. Amino group number and location on the molecule affect potency. For example, 2′,6′-diamino is more potent than 2′-amino. The 3-amino group of the deoxystreptamine can not be modified or activity is lost.

Like streptomycin the aminoglycosides also impair the permeability barrier posed by the cytoplasmic membrane allowing initial loss of smaller molecular weight materials followed by loss of larger molecular weight material. The reason for loss of permeability control is not finally understood but involves several factors. These include: an initial association of the positively charged aminoglycoside and anionic charges on the cytoplasmic (and outer membrane in gram-negative bacteria) membrane; an electron-transport dependent transfer of the agent across the cytoplasmic membrane dependent on a cross-membrane charge potential (interior negative), and binding at one or more sites to ribosomes involved in a ribosomal cycle. The probable result of these associations is the production of discontinuity in the membrane initially of the diameter of K^+ ions but later becoming large enough for amino acids, bases etc. and subsequently for proteins. It is also possible that synthesis of membrane proteins is preferentially prevented allowing membrane defects.

Kasugamycin selectively prevents binding of fMet-tRNA$_f$ to mRNA-30 initiation complexes. Spectinomycin, which is used mainly for therapy of *Neisseria gonorrhoeae* infections, is an aminocyclitol but not an aminoglycoside as it does not contain an aminosugar. High-level resistance to spectinomycin involves a mutation of the S5 protein of 30S ribosomal subunit. Kasugamycin and spectinomycin are not bactericidal, do not cause misreading and reversibly bind to the 30S subunit. Spectinomycin blocks ribosomes during initiation.

Inhibitors of factor G

Fusidic acid. Fusidic acid and related antibiotics are unsaturated carboxylic acids based on the cyclopentanoperhydrophenanthrane ring. Fusidic acid is active mainly on gram-positive bacteria with no significant activity on most gram-negative bacteria (although *Neisseria* sp. are exceptions). It well absorbed orally and is well distributed in most tissues but does not effectively enter cerebrospinal fluid. It has been used particularly for treatment of serious Staphylococcal infections under a variety of circumstances. It is metabolised by liver and excreted and concentrated in bile.

Fusidic acid is a selective inhibitor of elongation factor G in bacteria. It does so by stabilizing the ribosome-factor G guanosine diphosphate complex. This results in inhibition of factor G mediated GTP hydrolysis. It also inhibits binding of aminoacyl-tRNA to ribosomes. The result is inhibition of protein synthesis.

Enzymes of synthesis and reduction of dihydrofolate as targets

Trimethoprim and pyrimethamine

Trimethoprim and pyrimethamine are diaminopyrimidines which act as competitive inhibitors of dihydrofolate reductase (DHFR). This enzyme converts dihydrofolate to tetrahydrofolate which is used in one-carbon transfer in the synthesis of particularly thymidylate but also purines, methionine, thiamine, pantothenate and formyl methionyl tRNA. Trimethoprim has much higher affinity for most bacterial DHFR enzymes than the mammalian enzymes. About 50 000 times more trimethoprim is required to produce comparative inhibition of DHFR of mammalian cells as that for the DHFR of *E. coli*. Not all bacterial enzymes are this susceptible. For example, those of some anaerobic bacteria have significantly less affinity for trimethoprim and are more resistant to the drug.

Pyrimethamine is used as an antimalarial agent and for therapy of *Toxoplasma gondii* infections. It has 2 000-fold greater affinity for the malarial DHFR than for the mammalian enzyme. Folate deficiency is sometimes seen with therapy with pyrimethamine but almost never with trimethoprim. This is directly the result of the differential DHFR affinities noted in the preceding paragraph.

Trimethoprim is frequently used in combination with a sulfonamide such as sulfomethoxazole (co-trimoxazole). As such it has been widely used for pulmonary and urinary infections. This combination provides a 'double blockade' in the synthesis of tetrahydrofolate and is often synergistic. The actual occurrence of *in vivo* synergy is dependent on the relative concentrations of the two types of compounds in tissues and on the presence of inhibitors such as thymine or *p*-aminobenzoic acid. Co-trimoxazole is effective for treatment of *Pneumocystis carinii* and has been used in antibiotic prophylaxis. Its use in prophylaxis has not been associated with extensive resistance possibly due to the combination of the two agents.

Sulfonamides exert their antibacterial activity by reducing the rate of synthesis of functional dihydrofolate. Sulfonamides act as inhibitors of dihydropteroate synthetase (DHPS) by competing with *p*-aminobenzoic acid. It has been shown that sulfonamides act as alternative substrates to *p*-aminobenzoic acid to form a 7, 8-dihydropterin-sulfonamide adduct. However, the sulfonamide-dihydropterin adducts diffuse out of cells and do not reach concentrations high enough to inhibit enzymes of folate synthesis or to inhibit cell growth. Thus the effect of such adducts is to reduce the accumulation of the precursor pyrophosphate derivative of the reduced pteridine which would otherwise occur with inhibition of DHPS (see Fig. 3.4, Chapter 3). In this way there is less liklihood of reversal of the competitive inhibition through accumulation of one of the substrates.

The mechanism of the synergistic action of trimethoprim and sulfonamides is not yet fully understood. It is unlikely the sequential blockade of two steps in the synthesis of tetrahydrofolate accounts for synergy. Recently Poe has shown that sulfonamides bind to DHFR as well as trimethoprim. It has been proposed that the synergism results from increased inhibition of DHFR through binding of both inhibitors. However, sulfonamide concentrations needed to inhibit DHFR were above those obtained *in vivo* and would require a bacterial concentrating mechanism to obtain them. This view is supported by the observation that some bacteria resistant to sulfonamides due to a presumably insusceptible DHPS still show synergy with trimethoprim. The mechanism of synergistic activity remains unconfirmed at this time.

Miscellaneous targets

Metronidazole and nitrofurans

Metronidazole is active only on bacteria and cells that have anaerobic metabolism. This agent acts selectively on anaerobic organisms because they possess electron transfer components functioning at a sufficiently low redox potential (–430 mV or lower) to reduce the nitro group of metronidazole initiating the production of unstable intermediates which are biologically active on DNA. The final product of reduction does not seem active on DNA. Such components include ferrodoxin in bacteria and equivalent proteins in *Trichomonas vaginalis*. Uptake of the drug is closely associated with reduction resulting in a concentration gradient of unreduced metronidazole between the outside and inside of the cell.

Reduced metronidazole binds to DNA and causes strand breakage, dissolution of helix formation and degradation and simultaneously, directly or indirectly, inhibits the repair endonuclease DNase 1. The net result is the prevention of DNA replication and transcription.

Metronidazole inhibits the release of H_2 in the reaction in which pyruvate combines with phosphate producing acetyl phosphate, H_2 and CO_2. This seems so because metronidazole is reduced (accepts electrons) by an electron transport component which normally transfers electrons involved in the generation of H_2. Although the evolution of H_2 is inhibited, this is not the mechanism by which metronidazole kills cells, as noted in the preceding paragraph.

The capability of metronidazole to act as an electron sink for electron transport proteins and to affect DNA as described is probably a general property of 5-nitroimidazoles including tinidazole, dimetridazole, ornidazole and sulnidazole.

Nitrofurans probably act in a very similar manner to that of nitroimidazoles in that they act as an electron sink with the nitro group undergoing reduction producing biologically active intermediates. This leads in turn to the described effects on DNA. The extent of DNA damage varies among nitrofurans and is relatively less with nitrofurantoin than more strongly mutagenic nitrofurans. It is possible that inhibition of the conversion of pyruvate acetyl-Co-enzyme A may be relatively more important for this and similar agents.

A major difference does exist between nitroimidazoles and the nitrofurans in that the latter have redox potentials more positive than those of nitroimadazoles. Therefore, NADPH and NADH linked nitro-reductases of aerobic bacteria can reduce nitrofurans meaning nitrofurans are active on aerobic bacteria. The redox potential needed to reduce various nitrofurans differs.

Nitrofurans and metronidazole are mutagenic in bacteria. Nitrofurantoin, the most commonly used nitrofuran, is weakly mutagenic and is not known to be carcinogenic. Metronidazole has been reported to be carcinogenic but only at very high doses.

Nitrofurantoin is used almost exclusively as a urinary antiseptic. However, other nitrofurans have been used as topical (nitrofurazone, furazolidone (in the gut)), antitrypanosomal (nifurtimox), antifungal (nitrofuroxime), food additives (furylfuramide), additives in livestock feed (nitrovin) and as antitumor (nifuron) agents.

Metronidazole is used for therapy of anaerobic bacterial infections, *Trichomonas vaginalis, Giardia lamblia* and *En-*

tamoeba histolytica. It achieves excellent tissue concentrations including good concentrations in brain tissue.

Antituberculous drugs

Isoniazid (isonicotinic acid hydrazide) (INH). This agent is specifically active at low concentrations against certain mycobacteria including *M. tuberculosis.* It is relatively much less active on other bacteria. The mechanism of action of this agent is not known. Recently it has been proposed that INH reacts with tyrosine residues in protein thus inhibiting the action of certain proteins. It is known that INH causes production of yellow pigments by the 'Y-enzyme', decreases cell content of NAD^+ and inhibits mycolic acid synthesis. Several hypotheses have been advanced to explain these observations. Recently it has been suggested by Takayoma and Davidson that the pigment and the decrease of NAD^+ probably result from mycobacterial peroxidase acting on NAD^+, NADH, $NADP^+$ and NADPH in the presence of INH and oxygen. They further suggest a modified form of NAD^+ inhibits a NAD^+, $NADP^+$ linked enzyme (Δ^5-desaturase) involved in the synthesis of mycolic acid. These investigators suggest the reduction of mycolic acid results in impaired integrity of the cell envelope with ensuing loss of permeability control.

Ethambutol. The mechanism of action of this anti-mycobacterial drug is unknown. It may interfere with RNA metabolism.

Aminosalicylic acid (PAS). PAS inhibits mainly *Mycobacterium tuberculosis* and acts as an analog of *p*-aminobenzoic acid and is incorporated into folate analogs. This same mechanism may also inhibit synthesis of mycobactins used for iron transport.

Ethionamide. This agent is like INH in being a pyridine derivative and probably acts in a similar manner inhibiting mycolic acid synthesis.

Cycloserine. See inhibitors of the cell wall.

Inhibitors of protein synthesis. For streptomycin, kanamycin and amikacin–see aminoglycosides. Viomycin is a basic peptide and capreomycin is a cyclic peptide which probably act on the 30 S ribosomal subunit to inhibit protein synthesis.

Pyrazinamide. The mechanism of action is unknown.

Rifampicin. See nucleic acid inhibitors.

Selected references

Anderson, J.D. (1980). Fusidic acid: new opportunities with an old antibiotic. *Can. Med. Assoc. J.* **122,** 765–9.

Atherton, F.R., Hall, M.J., Hassall, C.H., Lambert, R.W., Lloyd, W.J. and Ringrose, P.S. (1979). Phosphonopeptides as antibacterial agents: mechanism of action of alaphosphin. *Antimicrob. Agents Chemother.* **15,** 696–705.

Bakker, E.P. (1979). Inophore antibiotics. In *Antibiotics,* vol. V, part 1, ed. F.E. Hahn, pp. 67–97. Springer-Verlag, Berlin.

Brown, J.R. and Ireland, D.S. (1978). Structural requirements for tetracycline activity. *Adv. Pharm. Chemother.* **15,** 161–203.

Buckel, P., Buchberger, A., Böck, A. and Wittmann, H.G. (1977). Alteration of ribosomal protein L 6 in mutants of *Escherichia coli* resistant to gentamicin. *Mol. Gen. Genet.* **158,** 47–54.

Chang, F.N. (1979). Lincomycin. In *Antibiotics,* vol. V, part 1, ed. F.E. Hahn, pp. 127–34. Springer-Verlag, Berlin.

Cozzarelli, N. (1977). The mechanism of action of inhibitors of DNA synthesis. *Ann. Rev. Biochem.* **46,** 641–68.

Davies, J. (1980). Aminoglycoside-aminocyclitol antibiotics and their modifying enzymes. In *Antibiotics in Laboratory Medicine,* ed. V. Lorian, pp. 474–89. Williams and Wilkins, Baltimore.

DeClercq, E., Descamps, J., Verhelst, Walterker, R.T., Jones, A.S., Torrence P.F. and Shugar, D. (1980). Comparative efficacy of antiherpes drugs against different strains of Herpes Simplex virus *J. Infectious Diseases* **141,** 563–74.

Downie, J.A., Gibson, F. and Cox, G.B. (1979). Membrane adenosine triphosphatases of prokaryotic cells. *Ann. Rev. Biochem.* **48,** 103–31.

Edwards, D.I. (1979). Mechanism of antimicrobial action of metronidazole. *J. Antimicrob. Chemother.* **5,** 499–502.

Feltham, S., Ronald, A.R. and Albritton, W.L. (1979). A comparison of the *in vitro* activity of rosamicin, erythromycin, clindamycin, metronidazole and ornidazole against *Haemophilus ducreyi* including β-lactamase producing strains. *J Antimicrob. Chemother.* **5,** 731–3.

Gale, E.F., Gundliffe, E., Reynolds, P.E., Richmond, M.H. and Waring, M.J. (1972). Inhibitors of nucleic acid synthesis. In *The Molecular Basis of Antibiotic Action,* pp. 173–267. John Wiley and Sons, London.

Ghuysen, J.M., Frère, J.M., Lehy-Bouille, M., Coyette, J., Dusart, J. and Nguyen-Distèche, M. (1979). Use of model enzymes in the determination of the mode of action of penicillins and Δ^3-cephalosporins. *Ann. Rev. Biochem.* **48,** 73–102.

Goldberg, I.H. and Friedman, P.A. (1971). Antibiotics and nucleic acids. *Ann. Rev. Biochem.* **40,** 775–810.
Grunert, R.R. (1979). Search for antiviral agents. *Ann. Rev. Microbiol.* **33,** 335–53.
Herman, R.P. and Weber, M.M. (1980). Isoniazid interaction with tyrosine as a possible mode of action of the drug in mycobacteria. *Antimicrob. Agents Chemother.* **17,** 170–8.
Hoeprich, P.D. (1978). Chemotherapy of systemic fungal diseases. *Ann. Rev. Pharmacol. Toxicol.* **18,** 205–31.
Kaji, A. and Ryoji, M. (1979). Tetracycline. In *Antibiotics,* vol. V, part 1, ed. F.E. Hahn, pp. 304–28. Springer-Verlag, Berlin.
Kitano, K. and Tomasz, A. (1979). *Escherichia coli* mutants tolerant to beta-lactam antibiotics. *J. Bacteriol.* **140,** 955–63.
Lacey, R.W. (1979). Mechanism of action of trimethoprim and sulfonamides: relevance to synergy *in vivo. J. Antimicrob. Chemother.* **5** (suppl. B), 75–84.
McCalla, D.R. (1979). Nitrofurans. In *Antibiotics,* vol. V, part 1, ed. F.E. Hahn, pp. 176–213. Springer-Verlag, Berlin.
McClatchy, J.K. (1980). Antituberculosis drugs: mechanisms of action, resistance, susceptibility testing and assays of activity in biological fluids. In *Antibiotics in Laboratory Medicine,* ed. V. Lorian, pp. 135–69. Williams and Wilkins, Baltimore.
Maren, T. (1976). Relations between structure and biological activity of sulfonamides. *Ann. Rev. Pharm. Toxicol.* **16,** 309–28.
Noguchi, H., Matsuhashi, M., Takaoka, M. and Mitsuhashi, S. (1978). New antipseudomonal penicillin, PC-904: affinity to penicillin-binding proteins and inhibition of the enzyme cross-linking peptidoglycan. *Antimicrob. Agents Chemother.* **14,** 617–24.
Nolan, C.M., Monson, T.P. and Ulmer, W.C. (1979). Rosaramicin versus penicillin G in experimental pneumococcal meningitis. *Antimicrob. Agents Chemother.* **16,** 776–80.
Norman, A.W., Spielvogel, A.M. and Wong, R.G. (1976). Polyene antibiotic-sterol interaction. *Adv. Lipid Res.* **14,** 127–70.
O'Callaghan, C.H. (1979). Description and classification of the newer cephalosporins and their relationships with the established compounds. *J. Antimicrob. Chemother.* **5,** 635–71.
Pestka, S. (1977). Inhibitors of protein synthesis. In *Molecular Mechanisms of Protein Biosynthesis,* ed. H. Weissbach and S. Pestka, pp. 468–555. Academic Press, New York.
Pestka, S. (1971). Inhibitors of ribosome functions. *Ann. Rev. Microbiol.* **25,** 487–562.
Poe, M. (1976). Antibacterial synergism: a proposal for chemotherapeutic potentiation between trimethoprim and sulfamethoxazole. *Science* **194,** 533–5.
Pongs, O. (1979). Chloramphenicol. In *Antibiotics,* vol. V, ed. F.E. Hahn, pp. 26–42. Springer-Verlag, Berlin.
Roland, S., Ferone, R., Harvey, R.J., Styles, V.L. and Morrison, R.W. (1979). The characteristics and significance of sulfonamides as substrates for *Escherichia coli* dihydropteroate synthase. *J. Biol. Chem.* **254,** 10337–45.
Sehgal, P.B. and Tamm, I. (1980). Antiviral agents: Determination of activity. In

Antibiotics in Laboratory Medicine, ed. V. Lorian, pp. 573–91. Williams and Wilkins, Baltimore.

Shoji, J. (1978). Recent chemical studies on peptide antibiotics from the genus Bacillus. *Adv. Appl. Microbiol.* **24,** 187–214.

Spratt, B.G. (1977). Penicillin binding proteins of *Escherichia coli:* general properties and characterization of mutants. In *Microbiology – 1977,* ed. D. Schlessinger, pp. 182–90. American Society Microbiology, Washington.

Spratt, B.G. (1977). Properties of the penicillin-binding proteins of *Escherichia coli* K 12. *European. J. Biochem.* **72,** 341–52.

Storm, D.R. and Toscano, W.A. (1979). Bacitracin. In *Antibiotics,* vol. V, part 1, ed. F.E. Hahn, pp. 1–17. Springer-Verlag, Berlin.

Storm, D.R., Rosenthal, K.S. and Swanson, P.E. (1977). Polymyxin and related peptide antibiotics. *Ann. Rev. Biochem.* **46,** 723–64.

Takayama, K. and Davidson, L.A. (1979). Isonicotinic acid hydrazide. In *Antibiotics,* vol. V, part 1, ed. F.E. Hahn, pp. 98–119. Springer-Verlag, Berlin.

Tanaka, N. (1975). Fusidic acid. In *Antibiotics,* vol. III, ed. J.W. Corcoran and F.E. Hahn, pp. 436–47. Springer-Verlag, Berlin.

Thammana, P. and Davies, J. (1975). Sensitivity and resistance to the antibiotic kasugamycin in *Escherichia coli.* In *Drug Action and Drug Resistance in Bacteria 2. Aminoglycoside Antibiotics* ed. S. Mitsuhashi, pp. 249–68. University of Tokyo Press, Tokyo.

Tipper, D.J. (1979). Mode of action of β-lactam antibiotics. *Reviews Infect. Dis.* **1,** 39–53.

Tomasz, A. (1979). The mechanism of the irreversible antimicrobial effects of penicillins: How the β-lactam antibiotics kill and lyse bacteria. *Ann. Rev. Microbiol.* **33,** 113–38.

Tomasz, A., Albino, A. and Zanati, E. (1970). Multiple antibiotic resistance in a bacterium with suppressed autolytic system. *Nature* **227,** 138–40.

Wallace, B.J., Tai, P.C. and Davis, B.D. (1979). Streptomycin and related antibiotics. In *Antibiotics,* vol. V, part 1, ed. F.E. Hahn pp. 272–303. Springer-Verlag, Berlin.

Welling, P.G. (1979). The esters of erythromycin. *J. Antimicrob. Chemother.* **5,** 633–4.

Whitley, R.J. and Alford, C.A. (1978). Developmental aspects of selected antiviral chemotherapeutic agents. *Ann. Rev. Microbiol.* **32,** 285–300.

Westley, J.W. (1977). Polyether antibiotics: Versatile carboxylic acid ionophores produced by Streptomyces. *Adv. Appl. Microbiol.* **22,** 177–225.

3

Mechanisms of resistance to antibacterial agents

A multiplicity of mechanisms of resistance exist for the various antibiotics. These include several situations where more than a single resistance mechanism exists for a single group of antibiotics. This chapter examines biochemical resistance mechanisms; their relationship to whole cells and to individual bacterial species is covered in Chapters 5 and 6. The genetics of resistance is provided in Chapter 4.

Resistance to β-lactam antibiotics

Associated with hydrolysis of the β-lactam ring

Enzymes capable of hydrolyzing β-lactam rings are termed β-lactamases. Many different β-lactamases exist among bacteria which vary in several properties including the types of β-lactams hydrolyzed. Considerable efforts have been directed to classification of β-lactamases of gram-negative bacteria in recent years. Properties frequently used to classify β-lactamases include: genetic basis (chromosomal v. plasmid); substrate profile (relative rates of hydrolysis of a range of β-lactams); susceptibility to inhibition by β-lactams (e.g. cloxacillin) and non β-lactams (e.g. *para*-chloromercuribenzoate); analytical isoelectric focusing, molecular weights and immunological cross-reactivity.

β-lactamases in gram-positive bacteria have not been extensively classified although enzymes of *Bacillus cereus, B. lichenformis* and *Staphylococcus aureus* have been extensively studied. One of the clinically most important mechanisms of β-lactamase specified resistance is among the staphylococci. β-lactamases of *S. aureus* and *S. epidermidis* are mainly active on benzylpenicillin, ampicillin, amoxycillin, phenoxymethylpenicillin and phenethicillin although some cephalosporins are hydrolyzed. For example, cephaloridine is much more rapidly hydrolyzed than cefazolin and cefazolin faster than cephalothin by β-lactamases of *S. aureus*. Methicillin, isoxazolyl penicillins and nafcillin are resistant to these β-lactamases. (However β-lactamases found in gram-negative bacteria do hydrolyze the penicillinase-resistant penicillins.) The semi-synthetic penicillins have a very low affinity for staphylococcal β-lactamase. For example, the K_m is 5×10^{-6} M for benzylpenicillin but 10^{-2} M for methicillin and cloxacillin.

Properties of β-lactamases from *Bacillus cereus, B. lichenformis* and *S. aureus* are given in Table 3.1. These enzymes are present as pre-penicillinases in the cytoplasmic membrane. In the case of the enzyme from *B. lichenformis* 749, there is known to be a processing proteolytic enzyme which cleaves a hydrophobic peptide of 24 amino acids from the N-terminus of the phospholipoprotein pre-penicillinase to release the extracellular penicillinase. It is probable that this is a general mechanism for release of

Table 3.1. *Some properties of penicillinases of gram-positive bacteria*

Source of enzyme	Genetic control	Extracellular penicillinase		
		%	mol. wt	mol. activity
B. cereus 569	inducible	90	30 600	1.6×10^5
B. lichenformis 749	inducible	50	29 000	1.1×10^5
S. aureus (PI258)	inducible	40	29 600	5×10^4

Source of enzyme	Amino acid			Cellular source	Genes
	number	cysteine	tryptophan		
S. aureus penI	260[a]	absent	absent	—	plasmid
B. lichenformis	270[a]	absent	present	phospholipoprotein pre-penicillinase of cell membrane with 24 additional amino acids and a phosphotidyl serine at the amino terminus	chromosomal

[a]40% of amino acid residues are the same, approximate number of amino acids.

β-lactamases; for gram-positive bacteria as extracellular enzymes and for gram-negative bacteria to the periplasm. Enzymes are synthesized with the hydrophobic leader sequence of amino acids to facilitate transport of the protein through the cytoplasmic membrane.

Most penicillinases of gram-positive bacteria are inducible whether chromosomally or plasmid specified. A model for control of penicillinases has been proposed by Imsande. The genes for control of synthesis of penicillinase are normally plasmid borne in *Staphylococcus* and chromosomal in *Bacillus lichenformis*. A somewhat simplified version of Imsande's model of penicillinase control is illustrated in Fig. 3.1. Inducers may be gratuitous in that they are able to turn on enzyme synthesis but are not hydrolyzed or are very slowly hydrolyzed. Examples are methicillin, cephalosporin C and 2-(2′ carboxyphenyl) benzoyl-6-aminopenicillanic acid (CBAP). Other inducers including benzylpenicillin are hydrolyzed. CBAP is preferable as an inducer as it is not hydrolyzed and is not an active antibiotic.

Fig. 3.1. Simplified model for regulation of penicillinase synthesis. penA gene is the structural gene for an anti-repressor protein which is inactive in the absence of inducer or presence of 5-methyltryptophan. In the presence of an inducer; the anti-repressor protein is active and combines with the repressor to inactivate it and allow synthesis of penicillinase by the penP gene. If the anti-repressor protein is inactive, the repressor is operative and prevents penicillinase synthesis.

Genes for control and synthesis – chromosomal, *B. lichenformis*, plasmid *S. aureus;* penA, structural gene for anti-repressor; penP, structural gene for pre-penicillinase, P, promoter, O, operator; penI, structural gene for repressor.

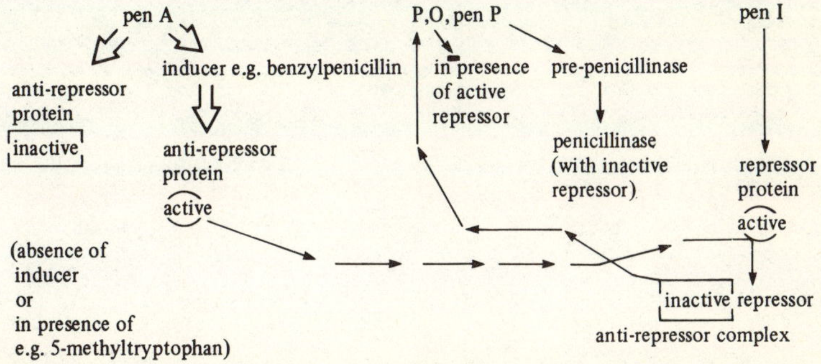

As noted in Chapter 5, penicillinases of gram-positive bacteria cause resistance mainly by detoxification through hydrolysis of the extracellular β-lactam by action of the cumulative β-lactamase produced by the whole population of β-lactamase producing cells.

It has been shown that some of the enzymes involved in synthesis of peptidoglycan may slowly release bound penicillin. The released penicillin is degraded. In one case the degradative reaction is that of β-lactam ring hydrolysis. Thus some synthetic enzymes are also β-lactamases, albeit relatively ineffectual ones. An example is the exocellular D, D-carboxypeptidase-transpeptidase of Streptomyces R39 in low ionic strength medium.

A variety of assays for β-lactamase activity have been described. Assays used include: spectrophotometric, hydroxylamine, microbiological, macro and micro iodometric, chromogenic cephalosporin substrate, and acidimetric/alkalimetric methods. Some of these are in use to screen clinical isolates for β-lactamase activity.

When many cephalosporins are acted on by β-lactamases, the reaction is less clearly defined than for penicillins (see Fig. 3.2).

The reaction shown in Fig. 3.2*b* varies depending on the cephalosporin undergoing attack. The number of equivalents of iodine reacting with penicilloic acid is 8; the number reacting with cephalosporin reaction products varies. Thus the iodometric assay is of very limited value for quantitative determination of cephalos-

Fig. 3.2. Action of β-lactamases on penicillins and cephalosporins.

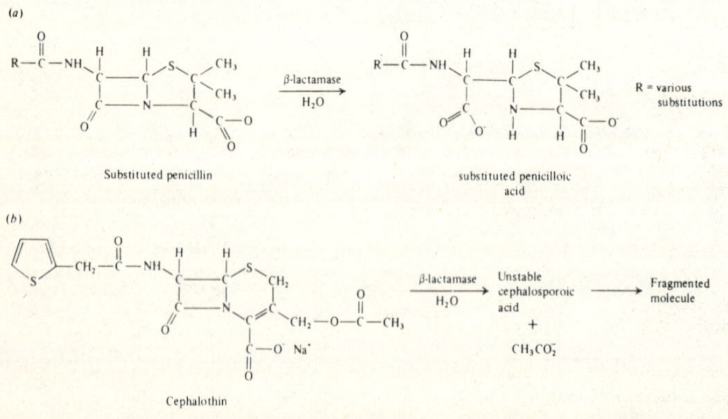

porinase activity with different cephalosporin substrates. The cephalosporin substrate nitrocefin changes color after β-lactam ring hydrolysis. It is an extremely sensitive assay method for all β-lactamases except TEM type (see Table 3.2) for which it is somewhat less sensitive.

The availability of the nitrocefin assay has shown that chromosomal β-lactamases are widespread, in fact, they may occur in all bacteria.

β-lactamases of gram-negative bacteria are classified broadly as: chromosomally mediated penicillinases, cephalosporinases or broad spectrum β-lactamases and R-plasmid mediated β-lactamases. Although this is a useful scheme for classification, absolute evidence for genetic origin of many β-lactamases is lacking.

Chromosomal penicillinases represent a small group of enzymes found mainly in strains of *Proteus mirabilis, P. morganii* and *P. thomasii.* The proof that all of these strains are chromosomal is not absolute and the possibility remains that some are R-plasmid specified. One enzyme, the Dalgleish enzyme originally included here is now known to be a plasmid specified enzyme (PSE-4).

The vast majority of chromosomal β-lactamases are cephalosporinases. In general they have molecular weights from 20 000 to 42 300 and isoelectric points from 3.9 to 8.6 (most are above pH 7.0). Most but not all are inhibited by cloxacillin and not inhibited by pCMB. These enzymes are classed as cephalosporinases because they hydrolyze selected cephalosporins (particularly cephaloridine) several times as rapidly as benzylpenicillin. They are either inducible or constitutive.

The chromosomal cephalosporinases are very wide spread among many types of bacteria. Some bacteria apparently produce more than one type of chromosomal β-lactamase including *Yersinia entercolitica* and *Pseudomonas cepacia.*

Chromosomal cephalosporinases are capable of producing resistance to cephalosporins and penicillins in bacteria which produce them. Studies have shown that inhibition of the chromosomal cephalosporinase found in *P. aeruginosa* (Sabath and Abraham enzyme) by cloxacillin decrease MICs of cephalosporin C. Mutants deficient in induction of the enzyme had lowered MICs of penicillin G, ampicillin and cephaloridine. Chromosomal β-lactamases play an important role in the intrinsic resistance of many bacteria to β-lactam antibiotics.

Many recently introduced cephalosporins are not hydrolyzed or hydrolyzed very poorly by some of the chromosomal cephalosporinases. This is particularly true of the 7-α-methoxy cephalosporins such as cefoxitin and cefmetazole. Resistance of a β-lactam to hydrolysis by a chromosomal cephalosporinase does not necessarily cause the bacterium to be susceptible to that agent. For example cefoxitin has poor or absent activity against many *Enterobacter* species and *P. aeruginosa* although their cephalosporinases do not hydrolyze cefoxitin or hydrolyze it very slowly. Rate of penetration of the gram-negative cell envelope by a β-lactam and affinity for penicillin-binding proteins are also major determinants of the activity of these agents. (See Chapter 5.)

Chromosomally specified broad spectrum β-lactamases are a

Table 3.2. *Properties of R-plasmid specified β-lactamases*

Designation	Rates of hydrolysis relative to benzylpenicillin (100)									Inhibition by		Molecular weight (10^3)
	Amp[a]	Car	Oxa	Met	Clo	Cep	Cpn	Cxn	Cfu	Clo	pCMB	
TEM-1	106	10	5	0	0	76	20	<10	<2	S	R	22
TEM-2	107	10	5	0	0	74	20	<10	<2	S	R	23.5
SHV-1	212	8	0	<2	<2	56	8	14	4	S	R/S[b]	17
HMS-1	253	14	<2	<2	2	183	3	<2	<2	S	S	21
OXA-1	382	30	197	332	190	30	15	0	<2	R	PS	23.3
OXA-2	197	15	646	23	200	37	25	<5	<2	R	R	44.6
OXA-3	178	10	336	29	350	44	10	<5	<2	R	R	41.2[c]
PSE-1	90	97	<2	<2	<2	18	<2	0	<2	R	S	28.5
PSE-2	267	121	317	803	371	32	<2	11	4	S	S	12.4
PSE-3	101	253			3	10		<1		S	R	12
PSE-4	88	150	8	16	<2	40	4	8	<2	R	R	32

[a]Abbreviations: Amp – ampicillin; Car – carbenicillin; Oxa – oxacillin; Met – methicillin; Clo – cloxacillin; Cep – cephaloridine; Cxn – Cephalexin; Cfu – cefuroxime.
[b]R – resistant; S – sensitive; R/S – resistant with benzylpenicillin as substrate but sensitive with cephaloridine as substrate; PS – partially sensitive.
[c]Probably exist as dimers.

small group of enzymes with similar activity on benzylpenicillin and cephaloridine. They occur principally in strains of various Klebsiella species although they are also found in *Enterobacter clocae* and *E. coli.* Except for the *E. coli* enzyme they are constitutive.

R-plasmid specified β-lactamases can be currently divided into 11 types based on isoelectric focusing and other properties noted above. Table 3.2 provides a summary of the properties of these enzymes. Fig. 3.3 provides the pattern of isoelectric focusing seen with the 11 enzymes.

TEM-1 β-lactamase is by far the most widely distributed of the enzymes representing more than the remainder of the enzymes combined in a study reported by Matthew.

The other enzymes were detected in the following decreasing order of frequency; TEM-2, OXA-1, OXA-2, SHV-1, OXA-3, PSE-4 with PSE-1,2,3, and HMS-1 in only 1 or 2 strains each.

The β-lactamase genes are known to be transposon-borne for TEM-1 and TEM-2 β-lactamases and may be for other β-lactamases. The world distribution of many of these β-lactamases has been summarized by Matthew and represented 25 countries,

Fig. 3.3. Isoelectric focusing patterns of plasmid-specified β-lactamases.

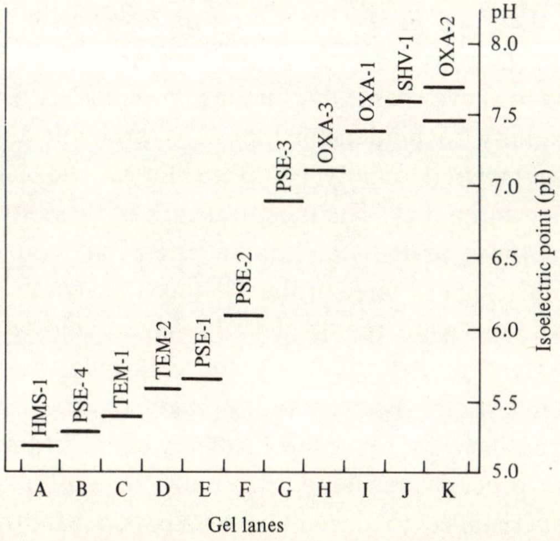

although organisms from large areas have not been examined. Of 363 gram-negative bacterial strains examined 77.4% had ß-lactamases belonging to the TEM group. The SHV-1 enzyme is unusual in being resistant to inhibition by *para*-chloromercuribenzoate with benzylpenicillin as substrate but susceptible to inhibition with cephaloridine as substrate. The amino acid sequence of TEM-1 and the nucleotide sequences of the gene coding for the enzyme have been determined.

Amino acid analysis of several ß-lactamases have shown that there is considerable overlap in structure among them. This suggests that there has been similar evolutionary origin of some of the genes for ß-lactamases.

Several ß-lactamases have been detected among strains of *Bacteroides fragilis* and other *Bacteroides* species. These enzymes are mainly cephalosporinases. At least five separate enzymes can be differentiated by isoelectric focusing. The amount of ß-lactamase produced varies among strains and, at least in some strains, is a major factor in resistance to ß-lactams.

Not associated with hydrolysis of the ß-lactam ring

Resistance to ß-lactam antibiotics without detectable ß-lactamase or acylase activity has been observed in many bacteria. Some strains have been derived as mutants selected within laboratories but others are naturally occurring strains.

Clinical isolates of *Streptococcus pneumoniae* with low-level resistance to ß-lactams (MIC of penicillin 0.16–1.6 μg/ml) have been shown to have penicillin-binding proteins (PBPs) with reduced capability to bind penicillin. It is probable that these proteins have lowered affinity for penicillin G, cloxacillin and probably other antibiotics. The most resistant of these strains had alterations of three of the four major PBPs found in sensitive strains. These changes are probably the effects of multiple mutations. Mutants with modified PBPs have been detected in laboratory-derived mutants. Tomasz and colleagues have shown that penicillin-resistant isolates of *S. pneumoniae* from South Africa acquired oxacillin (and thus probably penicillin) resistance in a stepwise process presumably by multiple mutation.

Low-level resistance to penicillin in *Neisseria gonorrhoeae* is associated with usually several genetic loci. Two of these loci which are associated with multiple drug resistance have changes in content of outer membrane protein (see Chapter 5).

The basis of methicillin resistance in staphylococci remains obscure. It is not due to methicillin inactivation but seems most likely to result from decreased affinity of penicillin-binding proteins for methicillin and penicillin. Expression of resistance depends on several factors. Resistance is usually enhanced at 30 °C compared to 37 °C and with 5% NaCl or 20% sucrose added to testing media. The chelating agent EDTA reduces resistance. Resistance is frequently expressed in only a minority of the population. The resistant proportion of the population is increased with NaCl or sucrose in the testing medium.

Tolerance to β-lactams represents the situation where the MBC (with the MBC defined as 99.9% killing) is much in excess of MIC. Definition of tolerance varies between investigators with the necessary ratio of MBC:MIC of penicillin after 18–24 hours growth defined as 10 by Best and 32 by Sabath. Tolerance must be differentiated from the so called Eagle effect where bactericidal activity is manifest for a small range of β-lactam concentrations but at higher concentrations such activity is reduced. Tolerance is observed naturally among certain streptococci, such as *Streptococcus sanguis*. Strains showing tolerance are likely deficient in autolytic activity either through a natural deficiency of autolysin or due to inhibition of that activity.

The cell wall of bacteria influences suceptibility levels for β-lactam antibiotics. Cell wall structure of *S. fecalis* has been considered to account for low resistance to penicillins and cephalosporins among these bacteria. However, recent investigations suggest that such resistance is most likely due to lessened affinity of penicillin-binding proteins for the β-lactams. Gram-negative bacteria pose a permeability barrier for some β-lactam antibiotics. The reason why this barrier varies among β-lactams and among gram-negative bacteria is not clear. Some β--lactams (e.g. cephaloridine and many cephalosporins in general) enter through the outer membrane pores of *E. coli* better than others (e.g. ampicillin and many penicillins). Outer membranes of some bacteria (*Hemophilus influenzae*) are more permeable to β-lactams than those of other bacteria (*Pseudomonas aeruginosa*). The effects of differential permeability account for differential susceptibility to various β-lactams, different MICs with identical β-lactamases in different strains and for inocula effects with some β-lactams. (See Chapters 2 and 5 for a more detailed discussion.)

R-factors can also cause differences in cell wall composition independent of β-lactamase activity.

Recently strains of *H. influenzae* resistant to ampicillin but lacking β-lactamase have been described. The mechanism of this resistance is currently unknown.

Mecillinam is generally more active on gram-negative Enterobacteriaceae than gram-positive bacteria. Richmond has shown that mecillinam acts like many cephalosporins in that it is not remarkably excluded by the envelope of *E. coli*. However, *P. aeruginosa* does exclude this drug and is resistant to it. Some β-lactamases (e.g. TEM) do hydrolyze mecillinam but have a high K_m for mecillinam. This means that relatively high and unphysiological concentrations of mecillinam are required before active hydrolysis is initiated. Thus to date β-lactamase mediated resistance has not been shown to be of importance for mecillinam.

Mecillinam activity is reduced by high inocula and with media of low osmolarity or conductivity. Phenotypic resistance is common and is associated with reduced growth rate. These cells revert to normal on withdrawal of the drug.

Stable mecillinam resistance has been detected in clinical specimens. An *E. coli* clinical isolate has been shown to be mucoid and to have a smaller cell mass and slower growth rate. In this mutant mecillinam induced spheres but septation was completed. Susceptible bacteria not completing septation became enlarged and were unstable.

Resistance to other agents acting on the cell wall

Cycloserine

Resistance to cycloserine may be due to the reduced efficiency of one or more transport systems for D-alanine and glycine and thus due to reduced transport of D-cycloserine. Some strains may alternatively exhibit elevated amounts of the target enzymes alanine racemase and D-alanine-D-alanine synthetase.

However it is probable that reduced transport is the more important mechanism.

Phosphonomycin

Resistance is most often due to defective transport systems for L-α-glycerophosphate or hexose phosphates. Resistance can also be due to mutations altering the affinity of pyruvate-UDP-*N*-acetylglucosamine transferase for phosphonomycin.

Bacitracin, vancomycin, ristocetin

Acquired resistance to these agents is uncommon although gram-negative bacteria are resistant. These relatively large antibiotics penetrate the pores of the outer membrane poorly and thus do not readily reach their targets in the cytoplasmic membrane.

Resistance to agents acting on the cytoplasmic membrane

Polymyxins

Several types of bacteria are naturally resistant to polymyxins including many gram-positive bacteria, *Proteus* species, *Serratia* species, *Providencia, Pseudomonas cepacia, Neisseria meningitidis* and *Neisseria gonorrhoeae.* Acquired resistance to polymyxins is very uncommon and in some instances is reversible when polymyxin is removed from the environment.

Strains of *Proteus* have been extensively examined and owe their resistance to the nature of their outer membranes. *P. mirabilis* spheroplasts are 400 times as sensitive to polymyxin B as whole cells. However this is not true for *P. cepacia* and *S. fecalis.*

Unstable mutants of *P. aeruginosa* resistant to polymyxin show a marked reduction in protein-LPS particles in the outer cell wall. Growth in low Mg^{2+} causes *P. aeruginosa* polymyxin resistance and is associated with a disorganization and crowding of protein-LPS particles in the outer membrane. Under these conditions there is a marked increase in the H_1 outer membrane protein. A polymyxin-resistant mutant of *Salmonella typhimurium* (*pmr*A) had lipopolysaccharide (LPS) which bound less polymyxin than the parent strain. This mutant has been proposed to have an alteration of the deep core or lipid A of its LPS.

In summary the mechanism (or mechanisms) of polymyxin resistance is not understood. Certain bacteria do not appear to bind polymyxin to either outer membrane or cytoplasmic membrane targets and others can effectively exclude the drug from the latter target.

Other agents

Most gram-negative bacteria are resistant to ionophores because their outer membranes exclude the agents from the cytoplasmic membrane. Most bacteria are resistant to polyenes because they lack sterols with which polyenes associate.

Resistance to agents acting on nucleic acids

DNA

Nalidixic acid acts on the nal subunit of DNA gyrase inhibiting DNA replication. This subunit or component of the gyrase is coded for by the *nal*A gene in *E. coli*. Resistance to this agent occurs frequently but such resistance is mutational. R-plasmid specified resistance has not been described. It is probable that most clinical resistance will be found to be from reduced affinity of the target *nal*A gene product for nalidixic acid.

Laboratory mutants produce single-step high-level resistance to nalidixic acid in most bacteria. These resemble the clinical isolates of resistant *E. coli* found in particularly urinary tract infections. Low-level resistant mutants can also be detected in laboratory and clinical isolates. These strains probably reduce the entry of nalidixic acid to the DNA gyrase.

Most gram-positive bacteria are resistant to nalidixic acid but the mechanism is unknown. Possibly these bacteria have an insensitive DNA gyrase.

Novobiocin susceptibility occurs in most gram-positive bacteria. Resistance in some gram-positives is associated with an increase of teichuronic acid probably decreasing wall permeability for the agent. Resistance in gram-negative bacteria is dependent on the structure of the outer membrane. Mutants super sensitive to novobiocin and other lipid soluble antibiotics have been described. Two of these from *E. coli* are *acr*A, with a 90% reduction in phosphate content of the lipid A region of LPS and *rfa*D with

D-glycero-D-mannoheptose replacing L-glycero-D-mannoheptose of LPS and a near absence of distal sugars of the LPS in the LPS core. Thus the low susceptibility of gram-negative bacteria is due to the LPS and outer membrane. These mutants also showed increased susceptibility to several other lipid soluble antibiotics including chloramphenicol and nalidixic acid.

Coumermycin A and novobiocin resistant mutants having a defective cou subunit of DNA gyrase have been detected.

RNA

Rifampicin resistance is not carried on R-factors although several plasmids cause bacteria to have increased susceptibility to the drug by an unknown mechanism. Mutational resistance to rifampicin is readily obtained among bacteria. However, clinical resistance has been uncommon except from treatment of meningococcal carriers. In the case of *M. tuberculosis* this is probably due to the use of combined antibacterial agents.

Resistance in most instances is due to a mutation altering an amino acid of the β-subunit of RNA polymerase. Most mutants are associated with single-step high resistance but some show intermediate levels of resistance.

Rifamicin has undergone considerable structural modification particularly at positions 3 and/or 4 of rifamycin SV. These modifications can enhance the activity of the agent on bacteria but they do not markedly change the effect on isolated RNA polymerase. Thus the enhancement appears due to better penetration through the cell envelope. The low susceptibility seen naturally in gram-negative bacteria is probably due to poor penetration. This apparently is a result of the large size of the molecule and thus poor penetration through membrane pores. The hydrophilic polysaccharide portion of the LPS molecule also presents a diffusion barrier to the molecule.

Resistance to agents acting on dihydrofolate-mediated functions

Several mechanisms of resistance exist for trimethoprim and sulfonamides. Acquired resistance to sulfonamides particularly among isolates from the urinary tract is now relatively common

(20–50%) especially among hospital inpatients. Acquired trimethoprim resistance remains less frequent with unusual exceptions (e.g. 52% resistance in *H. influenzae* from chronic respiratory disease in one report). Otherwise significant acquired resistance among staphylococci and gram-negative bacteria is usually only a few per cent of strains isolated. Both transferable and non-transferable resistance has been detected with long-term use of co-trimoxazole.

Plasmid-mediated resistance to sulfonamides has been shown for several bacteria to be due to the presence of a new dihydropteroate synthetase (DHPS). The new enzyme differs from the chromosomal enzyme in that it has reduced affinity for sulfonamides but not for *para*-aminobenzoic acid. In general the new enzyme differs in being more heat sensitive, and having a lower molecular weight (45 000 v. 49 000). It is not yet clear if all plasmid-specified DHPS enzymes are similar. Sulfonamide resistance is transposable so that a transposon could be widely distributed among plasmids.

It has also been reported that some chromosomally specified sulfonamide resistance is due to an altered DHPS. However, apparently these enzymes have a higher affinity for sulfonamides than the plasmid-specified enzymes and are inhibited by lower sulfonamide concentrations.

A similar mechanism of plasmid-specified trimethoprim resistance has also been described. At least two distinct types of 'plasmid' dihydrofolate reductases (DHFR) have been well characterized. As shown in Table 3.3 the enzymes require much higher concentrations of trimethoprim to cause 50% inhibition of DHFR activity than for the chromosome enzyme. Type II plasmid DHFRs is exceedingly insusceptible to trimethoprim and the other enzyme inhibitors noted in Table 3.3. Type I and II enzymes do not cross-react serologically. Both of these enzymes also have markedly reduced susceptibility to pyrimethamine and differ from the *E. coli* and *Citrobacter* chromosomal enzyme in several other respects. The R-factor enzymes also differ from the T-even phage and mammalian DHFRs. The DHFR specified by R-factor R388 has a subunit molecular weight of 10 500 and serologically cross-reacts with the type II enzymes. Tn402 enzyme also

Table 3.3. *Properties of dihydrofolate reductases from selected plasmids and bacterial strains*[a]

Source	Plasmid inc group	Mol. wt ($\times 10^3$)	Subunit mol. wt ($\times 10^3$)	pH optimum	T½ at 45 °C (min)	K_m for DHF[b] (μM)	K_i for TMP[b] (μM)	I_{50} (μM) TMP	AMP[b]	PYR[b]	Activity % NADH:NADPH
Chromosomal	—	20.5	18	7+	12.2	1–4	0.0027, 0.006	0.007,	0.004	0.002	8–16
Plasmid-type I	several	32–27	18	5.5	1.5–2.5	5–11	32–150	46+	4.4+	2.5+	0
Plasmid-type II	several	34–37	9	6.5	—	4–9	—	60 000	600+	378+	—

[a]See Bryan 1980 for references to source of data.
[b]Abbreviations: DHF – dihydrofolate; TMP – trimethoprim; AMP – amethopterin; Pyr – pyrimethamine.

cross-reacts with the type II enzyme and has 9000 molecular weight subunit. Thus both type II (Tn402) and type I (Tn7) enzymes reside on transposons. Amino acid sequencing of several DHFRs has been carried out.

The possession of an additional type of DHFR or DHPS represents a bypass mechanism of resistance. As shown in Fig. 3.4 the new enzyme allows the sites of inhibition by sulfonamides and trimethoprim to be bypassed and for tetrahydrofolate to be synthesized. This process also illustrates an important characteristic of plasmid-mediated resistance in that the resistance mechanism is dominant over the function normally inhibited by the antibiotic.

Other mechanisms of resistance have also been described for sulfonamides and trimethoprim. A reduction in accumulation of

Fig. 3.4. Sites of action and bypass mechanisms of plasmid mediated resistance for sulfonamides and trimethoprim.

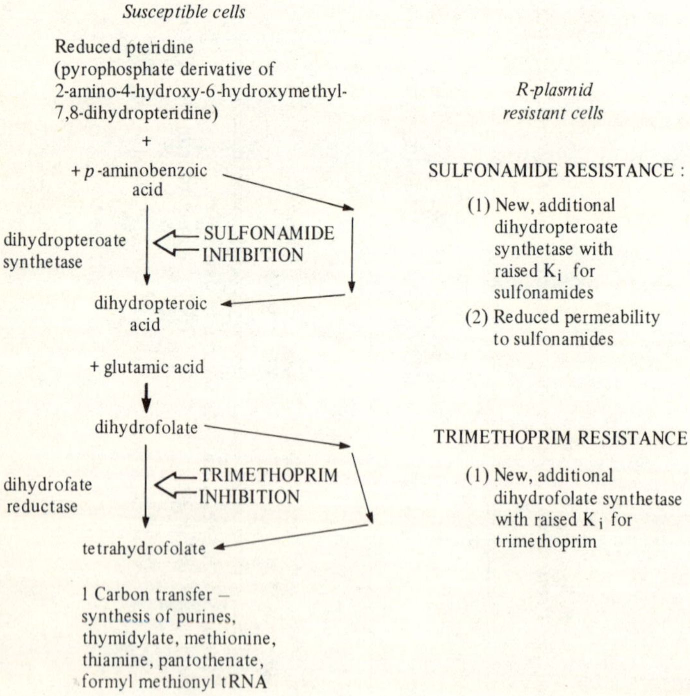

radiolabelled sulfonamide is sometimes associated with sulfonamide resistance. This is a possible mechanism of resistance but as drug accumulation can be altered by changes in other processes, it is unclear if impermeability is the primary mechanism. Some bacteria have been shown to have an increase in *para*-aminobenzoic acid concentration thus reducing the amount of folate analog synthesized.

Some bacteria which are resistant to sulfonamides and particularly some with lower resistance levels (<1000 μg/ml) do not have resistant DHPSs. The mechanism of resistance in these strains remains unclear.

Trimethoprim resistance also takes several additional forms. Many gram-negative bacteria have DHFRs which are inhibited by concentrations of trimethoprim much lower than that needed to inhibit growth of whole cells. Thus most bacteria but particularly gram-negative organisms have a permeability barrier to trimethoprim. In *P. aeruginosa* this permeability barrier is even greater producing MICs of 100 to 200 μg/ml, yet the DHFR is as susceptible as that of *E. coli* (MIC 0.01–1 μg/ml).

Some bacteria intrinsically resistant to trimethoprim like *Neisseria meningitidis, Clostridium perfringens* and *Bacteroides fragilis* have chromosomal DHFRs which have a sensitivity to trimethoprim 100 to 1000 times less than that of *E. coli.*

Bacteria which have a mutation affecting thymidylate synthetase require exogenous thymine or thymidine for DNA synthesis. These mutants are resistant to trimethoprim because the need for one carbon transfer from uridylate to thymidylate is not present. Such strains are not common but have been isolated from clinical materials under conditions where thymine or thymidine is present in the isolation medium.

Finally, an increased amount of DHFR has been described mainly in mutants resistant to trimethoprim although often selected for resistance to methotrexate. In some cases this has been combined with reduced sensitivity of the chromosomal DHFR to trimethoprim.

Resistance to trimethoprim is, like sulfonamide resistance, transposable and carried by several transposable elements.

Resistance to agents acting on ribosomes

Chloramphenicol

Acquired resistance to chloramphenicol is most frequently plasmid specified and can be transposable. Tn9 carries chloramphenicol resistance and is an unusual transposon, having a direct rather than an inverted repeat of the terminal DNA insertion sequence (ISI). The frequency and sites of transposition are more limited for Tn9 then many other transposons. Tn9 specifies a type I chloramphenicol acetyltransferase (see Table 3.4).

The commonest mechanism producing chloramphenicol resistance is the enzymatic acetylation of chloramphenicol (Fig. 3.5). The monoacetate is the initial product and in certain bacteria (*P. aeruginosa, S. aureus*) usually the only product. The diacetyl

Table 3.4. *Properties of chloramphenicol acetyltransferases from various bacteria*

Enzyme source	Inducible (I) or constitutive (C)	Monomer molecular weight ($\times 10^3$)	Sensitivity to thiol[a]	Immune class[b]
R-factor type I	C	22.5	S2+	A
R-factor type II	C	22.5	S5+	C
R-factor type III	C	24.5	S3+	B
Proteus mirabilis	C	22.5	S4+	A
Hemophilus parainfluenzae	C	22.5	S5+	C
Streptomyces acrimycini	C	24.5	S5+	C
Agrobacterium tumefaciens	I	24.5	S2+	C
Staphylococcus aureus				
type A	I	22.5	R	D
type B	I	22.5	R	D
type C	I	22.5	R	D
type D	I	22.5	R	D
Streptococcus pneumoniae	I	22.5	S1+	D
Streptococcus agalactiae	I	22.5	S2+	D
Streptococcus fecalis	I	22.5		D
Clostridium perfringens	I	22.5	S5+	

[a]S – sensitive; 1+, least sensitive; 5+, most sensitive; R, Resistant.

[b]A – reacts with antiserum to R-factor type I enzyme. B – reacts with antiserum to R-factor type III enzyme. C – no reaction with antisera to R-factor type I and III enzyme. D – reacts only with antiserum to Staphylococcal type C enzyme.

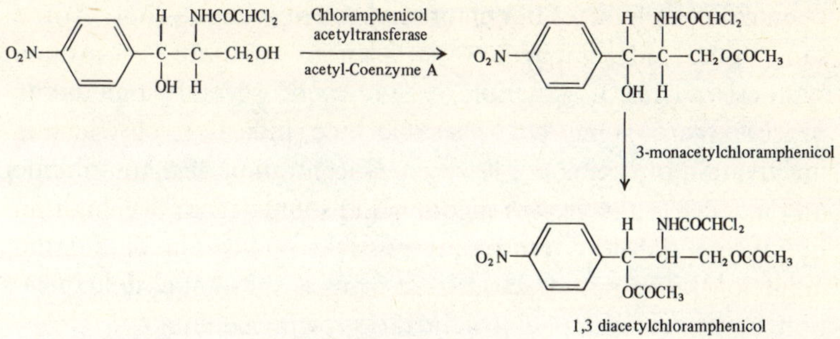

Fig. 3.5. Chloramphenicol acetylation.

derivative is normally produced by strains of *E. coli* and other Enterobacteriaceae. Assays commonly used to measure acetyltransferase activity include the use of acetyl-Coenzyme A radiolabelled in the acetyl moeity and a spectrophotometric assay (see Chapter 57, *Methods in Enzymology*, vol. XLIII, ed. J.H. Hash, Academic Press, New York).

Properties of chloramphenicol acetyltransferases from various bacteria are shown in Table 3.4. Other properties often reported are electrophoretic motility and affinity for 6-aminohexanoylchloramphenicol base. Most of the enzymes are plasmid mediated although there is some evidence the Proteus enzyme is chromosomal and for some bacteria the genetic basis is unknown (*Streptomyces, Agrobacterium, Clostridium*). The enzymes seem to be tetramers with pH optima about 7.8. Analysis of the amino acid sequence of the catalytic site of type I and type C enzymes has shown a very high degree of similarity despite the different bacterial origins of these two enzyme types. The active site sequence for the type I enzyme is histidine, alanine, valine, cysteine, asparagine, glycine, phenylalanine, histidine and valine. The type C enzyme is identical except tyrosine replaces phenylalanine and alanine replaces the terminal valine.

Chloramphenicol acetyltransferase results in antibiotic resistance for many bacteria by the inactivation of enough chloramphenicol in the growth medium to produce sub-inhibitory concentrations of that agent. This situation is true for most gram-negative bacteria, where the enzymes are generally constitutive. In those bacteria with inducible enzymes, the situation is more complex. Chloramphenicol is an inducer but chloramphenicol acetate is not. At the same time chloramphenicol is also an inhibitor of protein synthesis. Thus a situation exists where several competitive processes are going on. These include induction of enzyme, inactivation of antibiotic, lowered concentration of the antibiotic and inducer in the growth medium and some degree of inhibition of protein synthesis (and thus possible prevention of enzyme induction). The net effect in most cases is substantial detoxification of the chloramphenicol in the growth medium.

Infrequently among strains of *E. coli, Shigella* and *Salmonella* and commonly among strains of *P. aeruginosa* which possess R-factors no chloramphenicol acetyltransferase is present. It has been shown that an R^+ strain of *P. aeruginosa* did not have the transferase and that protein synthesis was inhibited by chloramphenicol. Mitsuhashi and his colleagues have shown reduced uptake of ^{14}C-chloramphenicol by R^+, enzyme negative strains. The suggestion from these studies is that R-factor specified impermeability for chloramphenicol is the mechanism of resistance. However, definite proof of impermeability is very difficult to provide. For example, a *P. aeruginosa* strain with chloramphenicol acetyltransferase shows reduced ^{14}C-chloramphenicol accumulation probably because acetylated chloramphenicol binds poorly to ribosomes.

Some R^+ strains of *P. aeruginosa* do produce chloramphenicol acetyltransferase but do not detoxify the medium as is found with *E. coli* strains, for example. Presumably these strains are resistant because little chloramphenicol reaches the cells periplasm and only that needs to be inactivated.

Some bacteria which do not have R-factors have low to medium levels of chloramphenicol resistance (e.g. *P. aeruginosa*) almost surely due to low permeability of the outer membrane. These strains accumulate ^{14}C-chloramphenicol poorly and are very much more sensitive to chloramphenicol as spheroplasts than whole cells.

E. coli mutants defective in the major outer membrane protein 1a are resistant to many antibiotics including chloramphenicol. Thus chloramphenicol appears to reach the periplasm mainly by outer membrane pores.

Macrolides and lincomycins

The commonest form of resistance to these agents is the erythromycin-inducible resistance to macrolides, lincosamides and streptogramin B-type (MLS) agents. This form of resistance, also referred to as dissociated resistance, is seen particularly in *Staphylococcus aureus, Streptococcus pyogenes* and *Streptococcus fecalis*. It is plasmid borne and appears to be carried on, at least, one type of transposon. The typical form of this resistance is that erythromycin acts at low concentrations (0.01–0.1 μg/ml) to induce resistance to all three classes of antibiotics. Other members of the group do not usually act as inducers although mutants which are lincomycin-inducible can be selected.

Constitutive mutants can be obtained which express resistance in the absence of an inducer. Patterns of resistance vary and can include generalized resistance or resistance to only some of the agents of the three groups.

In *S. pyogenes* constitutive generalized resistance requires two mutational events.

Resistance is due in the above strains to the presence of an enzyme activity which is normally inducible and is probably under control of a repressor protein. This enzymatic activity results in the methylation of probably two adenine residues in 23 S ribosomal RNA (rRNA). Reconstitution experiments using 23 S rRNA from induced and non-induced cells of *S. aureus* in a poly-U-directed polyphenylalanine-synthesis system demonstrated that resistance lay with that fraction. Methylated rRNA results in reduced ribosomal binding affinity for MLS type agents.

Interestingly *Streptomyces azureus* which produces thiostrepton methylates 23 S rRNA and is resistant to that agent.

Recently an erythromycin-inducible transposon, Tn917 has been described in *S. fecalis*. Erythromycin increases the frequency of erythromycin-resistant transconjugants. Tn917 is apparently induced to transpose.

Mutants resistant to erythromycin have been isolated following mutagenesis. These bind erythromycin poorly to 50 S ribosomes and contain an altered 50 S ribosomal protein termed the 50–8 protein.

Tetracyclines

Although investigation of resistance to tetracyclines has been very active, the mechanism remains unknown. Maximal levels of R-plasmid specified resistance are obtained by exposure of resistant cells to sub-inhibitory concentrations of tetracycline. (Plasmid pAM α 1 of *S. fecalis* is an exception and constitutively expresses full resistance.) Induction of resistance is associated with the synthesis of proteins which vary in number depending on the source of resistance genes. Plasmid pSC101 in *E. coli* specified three inducible proteins (34 000, 26 000 and 14 000), a constitutive protein (18 000) and a putative repressor (36 000). Plasmid R100-1 possessing the tetracycline transposon, Tn10 in *E. coli* causes induction of 15 000 and 36 000 molecular weight proteins. The 15 000 molecular weight protein is present in the inner membrane at about 4 000 molecules per membrane. The function of all of the proteins associated with Tn10 or pSC101 remains obscure.

The synthesis of proteins associated with pSC101 and Tn10 has been examined in minicells in which protein synthesis is directed only by the introduced plasmid DNA. The tet protein (36 000) of Levy and colleagues has been synthesized with cell-free extracts of *E. coli* and plasmid DNA. The latter studies were consistent with the presence of an inhibitor in extracts of uninduced cells.

Evidence to date is consistent with the view that in most bacteria studied, the tetracycline resistance system is under control of a repressor protein. Constitutively resistant mutants have been isolated for *E. coli* strains and in *Bacteroides fragilis*. In *B. fragilis* a sub-inhibitory amount of tetracycline induces both resistance and plasmid transferability. Mutants which are constitutive for both of these properties have been isolated. The 36 000 molecular weight protein of pSC101 has been tentatively identified as a repressor protein.

A 32 000 molecular weight protein has been associated with tetracycline resistance in *Staphylococcus aureus*. A decrease in amount of a 22 000 molecular weight protein has also been described. High-level tetracycline resistance in *S. aureus* is both plasmid and chromosomally mediated. However, both are associated with an inducible increase of resistance suggesting the mechanisms are similar.

Inducible tetracycline resistance is associated with a reduction in the rate of accumulation of tetracycline. In general reduction in uptake is usually about 2–5 fold whereas resistance increases are more often 10–100 fold. It remains unclear as to whether changes in accumulation are the sole cause of resistance or contribute with another mechanism to resistance or are secondary to an unknown resistance mechanism.

It has been shown in *E. coli* that cell protein synthesis directed by endogenous mRNA is more resistant to inhibition by tetracycline when prepared from resistant cells than from susceptible cells. It is therefore possible that tetracycline entering resistant cells less effectively interacts with the 30 S ribosomal subunit. This could be the result of the function of one of the proteins specified by the tetracycline resistant genes.

It also not clear if reduced tetracycline accumulation is due to lowered influx or enhanced efflux. Studies with an *E. coli* strain containing a plasmid having temperature-sensitive tetracycline resistance showed that efflux was similar at a temperature permitting resistance (30 °C) to a non-permissive temperature (42°C). However, recent studies on everted vesicles suggest increased efflux is important.

Recently the accumulation system for tetracycline has been shown to consist of rapid energy-independent and slow energy-dependent components in *E. coli* and *B. fragilis*. The slower component can be inhibited by cyanide, dinitrophenol and arsenate in *E. coli* and by rotenone in *B. fragilis*. When R222 is introduced into *E. coli* the characteristics of the accumulation process change particularly by being insensitive to dinitrophenol and by showing a reduced rate of initial rapid uptake. These changes indicate that the transport system in resistant cells is not the same as that in sensitive cells. It is likely an active transport carrier (tet protein?) is derepressed to give energy-dependent efflux. *B. fragils* R^+ cells which are induced or constitutive for resistance no longer show the energy-independent slow uptake but still exhibit the rapid energy-independent phase. Thus the changes in tetracycline accumulation patterns in R^+ *E. coli* and *B. fragilis* differ significantly in the effects produced on the initial energy-independent uptake. However, Tait *et al.* have shown that plasmids pSC101 and pMB9 in *E. coli* differ in that the former markedly affects the initial energy-independent tetracycline flux whereas pMB9 does not. However, both plasmids prevent slow energy-dependent uptake. The last effect seems a common feature in all systems studied to date.

Two types of tetracycline resistance occur among plasmids in *E. coli.* Plasmid RIP 111 specifies *tet*A type resistance and RIP69 the *tet*B type. The former shows no increase of resistance to minocycline or β-chelocardin, a tetracycline-like antibiotic. Furthermore sub-inhibitory tetracycline does not induce resistance to chelocardin. In the *tet*B type of pattern, elevated resistance to minocycline occurs and sub-inhibitory tetracycline induces resistance to tetracycline and β-chelocardin. β-chelocardin is not an inducer and selects for constitutively resistant mutants. Chabbert and Scavizzi have pointed out the similarity of the *tet*B pattern to inducible erythromycin resistance and suggested this resistance may be due to alteration of the 30 S ribosomal subunit.

A tetracycline-resistant mutant in *E. coli* has been shown to possess an altered small ribosomal protein. The mutation maps in the ribosomal protein gene region of the chromosome.

Table 3.5. *Mechanisms of resistance to aminoglycoside and aminocyclitol antibiotics*

Mechanism		
(A) Enzymic modification	Ring	Resistance may be specified to[a]
(1) *N*-acetylation		
AAC(6′)[b]	aminohexose I	KmA,Am,KmB, DDKmB,Tm,Si, Ne,Nm,Bu,Ri
AAC(2′)	aminohexose I	Gm,Si,Tm,Bu,Li
AAC(3)[b]	deoxystreptamine	Gm,Si,(KmA,KMB, DDKmB,Tm)
(2) *O*-phosphorylation		
APH(3′)[b]	aminohexose I	KmA,KmB,Nm
APH(2″)	aminohexose III	Gm,Si,Ne
APH(5″)	ribose	Li,Ri
APH(3″)	glucosamine	Sm
APH(6)	streptidine	Sm

(3) *O*-nucleotidylation		
ANT(3″)(9)	3″-glucosamine (Sm) 9-spectinomycin	Sm,Sp
ANT(2″)	aminohexose III	KmA,KmB,DDKmB, Tm,Gm,Si
ANT(4′)	aminohexose I	KmA,KmB,Am,Tm, Nm,Li,Bu,Ri
ANT(6)		Sm
ANT(9)	spectinomycin	Sp

(B) Impermeability or lack of transport
- (1) Most anaerobic bacteria – bacteria unable to carry out oxygen or nitrate respiration
- (2) Fermentative bacteria with absent or limited respiration (example streptococci)
- (3) Small colony phenotypes of many bacteria with impaired terminal respiration or proton leaky
- (4) R-factor specified – apparent (unproven) impermeability
- (5) Wide resistance to aminoglycosides due to apparent impermeability – especially *Pseudomonas*
- (6) Mutations of the gram-negative bacterial cell envelope – protein 1a of *E. coli* outer membrane

(C) Ribosomal
- (a) protein – high-level resistance. Streptomycin (S12), spectinomycin (S5)
  – low level or in combination with membrane energization mutants – deoxystreptamine group. Examples S6 – kanamycin, L6 – gentamicin
- (b) RNA – kasugamycin

[a]Low-level resistance may be specified to additional aminoglycosides. Abbreviations: KmA – kanamycin A; Am – amikacin; KmB – kanamycin B; DDKmB – dideoxykanamycin B; Tm – tobramycin; Gm – gentamicins; Si – sisomicin; Ne – netilmicin; Nm – neomycin B; Bu – butirosin; Ri – ribostamycin; Li – lividomycin; Sm – streptomycin; Sp – spectinomycin.

[b]Enzyme variants of these groups exist with different substrate affinities and can produce different resistance patterns. A wide pattern is shown for AAC(6′) and a narrow and wide (resistances in brackets) for AAC(3). AAC – aminoglycoside acetyltransferase; APH – aminoglycoside phorphoryl transferase; ANT – aminoglycoside nucleotidyl transferase. Number in brackets indicates ring position modified.

Aminoglycosides

Table 3.5 list mechanisms of resistance to aminoglycosides. Of these the commonest among clinical isolates of bacteria are enzymic mechanisms and those associated with lack of transport in anaerobic bacteria. The basis of anaerobic resistance is discussed in Chapter 5 in detail.

Most enzymic resistance is specified by plasmid DNA and some are known to be carried by transposons (see Chapter 4). As noted in Table 3.5, enzymes capable of amino acetylation (AAC), hydroxyl phosphorylation (APH) or hydroxyl nucleotidylation (ANT) of various rings of streptidine and deoxystreptamine aminoglycosides have been described. The abbreviations given in Table 3.5 for each enzyme activity are widely accepted. The number in brackets refers to the site of the amino or hydroxyl group modified.

There is a clear correlation between the possession of enzymes capable of modifying aminoglycosides and aminoglycoside resistance. Resistance is achieved through a rate competition mechanism as originally proposed by Bryan and colleagues. The mechanism by which inactivating enzymes achieve resistance in whole cells is considered in detail in Chapter 5. In summary, enzymic inactivation need only take place for that aminoglycoside that undergoes cell transport. This represents only a small fraction of the total aminoglycoside in the medium. The inactivating enzyme must have a high enough affinity (i.e. a low K_m) to initiate inactivation at or below those concentrations where transport is initiated. The transported drug is modified and does not bind to the ribosomal target resulting in resistance to the drug. Resistance therefore depends in any given circumstance on the relative rates of inactivation and transport. Inactivation rates depend especially on the K_m for aminoglycoside substrate and also on the V_{max} and total enzyme present. Transport rates depend on cell energy source, concentration and presence of transport antagonists like divalent cations and low pH and periplasmic drug concentrations in gram-negative bacteria.

Small colony variants of several types of bacteria have been isolated which are resistant to deoxystreptamine aminoglycosides.

Some have been detected in clinical infections. These organisms usually have a reduced growth rate and have abnormalities of terminal cytochromes, respiratory quinones or are proton leaky. These are due to one or more mutations (see Chapter 4, Table 4.1). The transport rate of aminoglycosides is reduced due to reduced energization of the drug transporter in the membrane or to a decrease in the membrane potential of the cell (see Chapter 5).

A few examples of aminoglycoside resistance specified by R-plasmids but not associated with enzymic inactivation have been described. These resistances are unlike most permeability resistance in that they are to a limited spectrum of aminoglycosides (e.g. kR102, streptomycin and kanamycin) rather than to the group as a whole. R^+ strains showed decreased accumulation but the reason for this is unknown.

P. aeruginosa strains detected as resistant to gentamicin, amikacin or tobramycin mostly from nosocomial infections do not have inactivating enzymes and show increased resistance to all tested aminoglycosides. They are resistant due to apparent impermeability to aminoglycosides. Strains of *P. aeruginosa* from burn patients or from some urinary tract infections often possess inactivating enzymes. The enzyme-containing strains usually are associated with higher resistance levels. They are probably selected for by the very high aminoglycoside concentrations obtained by topical use in burns or in urinary infections.

The common low-level broad aminoglycoside-resistant nosocomial strains may have an altered outer membrane protein pattern on gel electrophoresis. The relationship of this change to aminoglycoside resistance is unknown.

Some strains of *E. coli* lacking outer membrane protein 1a are resistant to aminoglycosides as well as other antibiotics. Strains with ribosomal mutations resulting in high level streptomycin resistance are occasionally detected among clinical isolates of bacteria. Strains of *N. gonorrhoeae* have been isolated showing ribosomal resistance to spectinomycin or streptomycin. Laboratory derived mutants affecting ribosomal proteins are noted in Table 3.5.

Fusidic acid

Resistance to fusidic acid can be detected at a frequency of 10^{-5}–10^{-6} per cell generation *in vitro* with *S. aureus*. Resistance following treatment is not as frequent as expected in view of the high frequency of mutational resistance. However, fusidic acid is commonly used with a second anti-staphylococcal agent especially a penicillin. Resistance to fusidic acid and penicillinase production has been detected on some staphylococcal plasmids.

Clinical isolates of *S. aureus* resistant to fusidic acid may be due to mutation or associated with a plasmid. Mutational resistance in *E. coli* and probably *S. aureus* is due to modification of elongation factor G causing a reduced affinity for fusidic acid.

Plasmid-mediated resistance occurs in *S. aureus* and is also found on several R-factors of gram-negative bacteria. A study of a sensitive mutant of an *E. coli* strain which is normally resistant to fusidic acid showed changes in the lipid composition of the cell envelope. The sensitive strain had reduced amounts of phosphatidylethanolamine and cyclopropane fatty acids compared to the intrinsically resistant wild type *E. coli* K12. The susceptible mutant apparently had a defective cyclopropane fatty acid synthetase. Introduction of R-factors specifying fusidic acid resistance into this strain increased the phosphatidylethanolamine and cyclopropane fatty acid content of the envelope similar to that of the wild type resistant strain. The suggestion is that these R-factors specify enzymes for cyclopropane fatty acid and phosphatidylethanolamine synthesis. In turn the changes in the envelope apparently prevent fusidic acid entry. The proposal is attractive and consistent with results to date but requires further confirmation of the relationship of drug entry and cell envelope lipids. Of interest the sensitive mutant also had much increased susceptibility to erythromycin and penicillin G.

In *S. aureus* plasmid-specified resistance is not associated with drug inactivation nor a modified elongation factor G. A strain carrying plasmid genes for fusidic acid resistance showed a reduced ratio of phosphatidylglycerol to lysylphosphatidylglycerol relative to a nearly isogenic strain not having fusidic acid resistance. Enhanced resistance occurs rapidly on incubation of plasmid-containing strains with fusidic acid. Following this procedure cells showed a pronounced decline in the same phospholipid ratio. These findings again suggest that plasmids may modify membrane lipids and perhaps reduce fusidic acid entry. Further clarification of this interesting relationship is warranted.

Miscellaneous targets

Metronidazole

Naturally occurring resistance to metronidazole among most clinically important bacteria is very uncommon. Intrinsic resistance to the agent occurs among aerobic and facultative bacteria which do not possess electron transport components functioning at a sufficiently low redox potential to reduce metronidazole. It is a reduced intermediate which is the active species of the agent.

Bacteroides fragilis strains resistant to metronidazole obtained by mutation have been shown to have reduced levels of pyruvate dehydrogenase activity. A resistant clinical isolate also had this finding. This resulted in secondary changes in the end products of glucose metabolism and in reduced uptake of the drug. Pyruvate dehydrogenase participates in the phosphoroclastic reaction of anaerobic bacteria and protozoa. It results in the generation of reducing equivalents normally used to produce H_2. In the presence of metronidazole, the drug is reduced. The deficiency of pyruvate dehydrogenase apparently decreases the capability to reduce metronidazole and generate the active intermediate.

Nitrofurans

Mutations resistant to nitrofurans can be readily obtained in *E. coli* and *S. aureus*. Mutants lack an O_2-insensitive NADPH linked reductase responsible for nitrofuran reduction. These mutants are susceptible to nitrofurans anaerobically as they still have an O_2-sensitive NADH linked enzyme active only anaerobically. In the absence of reduction, active intermediates are not generated.

Naturally occurring resistant clinical-isolates are uncommon. It is likely that mutants would need to lack both reductases. Plasmid-mediated low-level resistance has been detected and was associated with a reduced rate of drug reduction. In general resistance has not been a problem in treatment of urinary tract infections.

Antituberculous drugs

Isoniazid (INH). Resistant cells of *M. tuberculosis* incorporate reduced amounts of INH which normally enters cells by an active process apparently dependent on respiratory energy.

p-Aminosalicyclic acid (PAS). Resistance is probably due to a new dihydropteroate synthetase with reduced affinity for PAS.

Ethambutol and Ethionamide. The mechanism of resistance is unknown for these agents.

Inhibitors of protein synthesis. A variety of cross-resistance patterns occurs for streptomycin, kanamycin, viomycin and capreomycin in *M. tuberculosis.* It is clear that high-level ribosomal resistance to streptomycin occurs in *M. tuberculosis.* Evidence for plasmid-mediated resistance to streptomycin and perhaps to kanamycin and capreomycin has also been reported. Viomycin-capreomycin resistance may occur as a result of mutations affecting in one case the 50 S ribosomal subunit and in another, the 30 S subunit. Low-level kanamycin resistance may be due to a ribosomal mutation. Cross-resistance patterns may take at least six forms. The relationship among the agents requires susceptibility testing.

Pyrazinamide. Resistant strains of *M. tuberculosis* cannot convert pyrazinamide to pyrazinoic acid (PAO) by pyrazinamidase activity, unlike susceptible strains. PAO may be the active component.

Resistance to heavy metals

Resistance to heavy metals is a frequent phenotype possessed by many plasmids. The best studied mechanism of resistance is that to mercury.

In those situations examined to date, the mercuric ion (Hg^{2+}) is reduced to volatile Hg^0 which is lost from the growth medium. Reduction of Hg^{2+} invloves flavoproteins and requires a thiol reagent. NADPH is the preferred electron donor. Resistance is inducible. Interestingly plasmids specifying mercuric reductase activity also specify an inducible mercuric ion transport system. The reductase (merA) and transport system (merT) genes may form an inducible operon as part of the mer region of plasmids. Mercury resistance is found on transposons (e.g. Tn501).

Several organomercurials such as phenylmercuric, methylmercuric and ethylmercuric are cleaved by organomercurial lyases to release the organic compound and Hg^{2+}. The latter is subsequently reduced. Resistance to other organomercurials like *p*-hydroxymercuribenzoate, merbromin and fluoresceinmercuric acetate is not associated with significant enzymatic cleavage. Resistance in the last case has been speculated to be due to impairment of permeability.

The spectrum of resistance varies among plasmids and bacteria. Narrow spectrum resistance to Hg^{2+}, merbromin and fluoresceinmercuric acetate is seen with plasmids in *E. coli* and in *P. aeruginosa* (where additional low resistance to *p*-hydroxymercuribenzoate is also present). Broad spectrum plasmid resistance includes phenylmercuric acetate and thimerosal in *E. coli* and in *P. aeruginosa,* additionally, methyl and ethyl mercuric compounds. *S. aureus* plasmids show a broad spectrum pattern without resistance to merbromin.

Plasmid-mediated cadmium resistance in *S. aureus* is constitutive and associated with an acquired failure of the energy-dependent manganese system to transport cadmium as an alternative substrate. Resistance is therefore due to a failure of cadmium to enter the cell.

Resistance to lead, tellurium, silver, zinc, nickel, arsenate, arsenite, bismuth, antimony and cobalt is also specified by plasmids. Selenite resistance is also found but it is not clearly plasmid specified. Resistance to arsenate may be due to reduced affinity of an inorganic phosphate transport system for arsenate. The mechanisms of resistance to the other agents are unknown.

Selected References

Anderson, J.D. (1977). Mecillinam resistance in clinical practice – a review. *J. Antimicrob. Chemother.* **3** (suppl. B), 89–96.

Barbour, A.G. and Mayer, L.W. (1980). Mecillinam resistance and small cell volume: the *in vivo* selection of an *Escherichia coli* mutant. In *Current Chemotherapy and Infectious Disease,* ed. J.D. Nelson and C. grossi, pp. 715–16. American Society for Microbiology, Washington.

Britz, M.L. and Wilkinson, R.G. (1979). Isolation and properties of metronidazole-resistant mutants of *Bacteroides fragilis. Antimicrob. Agents Chemother.* **16,** 19–27.

Bryan, L.E. (1980). Mechanisms of plasmid mediated drug resistance. In *Plasmids and Transposons,* ed. C. Stuttard and K.R. Rozee, pp. 57–81. Academic Press, New York.

Bryan, L.E. (1979). Resistance to antimicrobial agents: the general nature of the problem and the basis of resistance. In *Pseudomonas aeruginosa. Clinical Manifestations of Infection and Current Therapy,* ed. R.G. Doggett, pp. 219–71. Academic Press, New York.

Buchanan, C.E. (1977). Altered proteins in penicillin-resistant mutants of *Bacillus subtilis.* In *Microbiology 1977,* ed. D. Schlessinger, pp. 191–4. American Society for Microbiology, Washington.

Chopra, I. and Howe, T.G.B. (1978). Bacterial resistance to the tetracyclines. *Microbiol. Rev.* **42,** 707–24,

Chopra, I. (1976). Mechanisms of resistance to fusidic acid in *Staphylococcus aureus. J. Gen. Microbiol.* **96,** 229–38.

Coleman, W.G. and Lieve, L. (1979). Two mutations which affect the barrier function of the *Escherichia coli* K–12 outer membrane. *J. Bacteriol.* **139,** 899–910.

Datta, N., Nugent, M., Amyes, S.G.B. and McNeilly, P. (1979). Multiple mechanisms of trimethoprim resistance in strains of *Escherichia coli* from a patient treated with long term co-trimoxazole. *J. Antimicrob. Chemother.* **5,** 399–406.

Davies, J. (1980). Aminoglycoside-aminocyclitol antibiotics and their modifying enzymes. In *Antibiotics in Laboratory Medicine,* ed. V. Lorian pp. 474–89. Williams and Wilkins, Baltimore.

Davies, J. and Smith, D.I. (1978). Plasmid-determined resistance to antimicrobial agents. *Ann. Rev. Microbiol.* **32,** 469–518.

Fayolle, F., Privitera, G. and Sebald, M. (1980). Tetracycline transport in *Bacteroides fragilis. Antimibrob. Agents. Chemother.* **18,** 502–5.

Fitton, J.E., Packman, L.E., Harford, S., Zaidenzaig, Y. and Shaw, W.V. (1978). Plasmids and the evolution of chloramphenicol resistance. In *Microbiology 1978,* ed. D. Schlessinger, pp. 249–52. American Society for Microbiology, Washington.

Fling, M.E. and Elwell, L.P. (1980). Protein expression in *Escherichia coli* minicells containing recombinant plasmids specifying trimethoprim-resistant dihydrofolate reductases. *J. Bacteriol.* **141,** 779–85.

Foster, T.J., Nakahara, H., Weiss, A.A. and Silver, S. (1979). Transposon A-generated mutations in the merucuric resistance genes of plasmid R100–1. *J. Bacteriol.* **140,** 167–81.

Foulds, J. and Chai, T.J. (1978). New major outer membrane protein found in an *Escherichia coli tol* F mutant resistant to bacteriophage Tulb. *J. Bacteriol.* **133,** 1478–83.

Ghuysen, J.M., Frère, J.M., Leyh-Boiulle, M., Coyette, J. Dusart, J. and

Nguyen-Distèche, M. (1979). Use of model enzymes in the determination of the mode of action of penicillins and Δ^3-cephalosporins. *Ann. Rev. Biochem.* **48,** 73–102.

Hamilton-Miller, J.M.T. (1979). Mechanisms and distributions of bacterial resistance to diaminopyrimidines and sulfonamides. *J. Antimicrobial. Chemother.* **5**(suppl B), 61–74.

Hamilton-Miller, J.M.T. and Smith, J.T. (eds.) (1979). *Beta-Lactamases.* Academic Press, London.

Horne, D. and Tomasz, A. (1980). Lethal effects of a heterologous murein hydrolase on penicillin-treated *Streptococcus sanguis. Antimicrob. Agents Chemother.* **17,** 235–46.

Imsande, J. (1978). Genetic regulation of penicillinase synthesis in gram-positive bacteria. *Microbiol. Reviews* **42,** 67–83.

Kaji, A. and Ryoji, M. (1978). Tetracycline. In *Antibiotics,* vol. V, part 1, ed. F.E. Hahn, pp. 304–28. Springer-Verlag, Berlin.

Kenward, M.A., Brown, M.R.W., Hesslewood, S.R. and Dillon, C. (1978). Influence of R-plasmid RP1 of *Pseudomonas aeruginosa* on cell wall composition, drug resistance and sensitivity to cold shock. *Antimicrob. Agents Chemother.* **13,** 446–53.

Kono, M. and O'hara, K. (1976). Mechanism of chloramphenicol-resistance mediated by KP102 factor in *Pseudomonas aeruginosa. J. Antibiotics.* **29,** 176–80.

Krogstad, D.J. and Moellering, R.C. (1980). Combinations of antibiotics, mechanisms of interaction against bacteria. In *Antibiotics in Laboratory Medicine,* ed. V. Lorian, pp. 298–341. Williams and Wilkins, Baltimore, London.

Lacey, R.W. (1979). Mechanism of action of trimethoprim and sulfonamides: relevance to synergy *in vivo. J. Antimicrobiol. Chemother.* **5** (suppl. B), 75–84.

Leung, T. and J.D. Williams. (1978). β-lactamases of subspecies of *Bacteroides fragilis. J. Antimicrob. Chemother.* **4** (suppl. B), 47–54.

Levy, S.B. and McMurry, L. (1978). Probing the expression of plasmid-mediated tetracycline resistance in *Escherichia coli.* In *Microbiology 1978,* ed. D. Schlessinger, pp. 177–180. American Society for Microbiology, Washington.

McCalla, D.R. (1979). Nitrofurans. In *Antibiotics,* vol. V, part 1, ed. F.E. Hahn pp. 176–213. Springer-Verlag, Berlin.

McClatchy, J.K. (1980). Antituberculous drugs: mechanisms of action, drug resistance, susceptibility testing and assays of activity in biological fluid. In *Antibiotics in Laboratory Medicine,* ed. V. Lorian, pp. 135–69. Williams and Wilkins, Baltimore, London.

McMurry, L. and Levy, S.B. (1978). Two transport systems for tetracycline in sensitive *Escherichia coli:* Critical role for an initial rapid uptake system insensitive to energy inhibitors. *Antimicrob. Agents Chemother.* **14,** 201–9.

Malke, H. (1978). Zonal-pattern resistance to lincomycin in *Streptococcus pyogenes:* genetic and physical studies. In *Microbiology 1978,* ed. D. Schlessinger, pp. 142–5. American Society for Microbiology, Washington.

Manniello, J.M., Heymann, H. and Adair, F.W. (1978). Resistance of spheroplasts and whole cell of *Pseudomonas cepacia* to polymyxin B. *Antimibrob. Agents Chemother.* **14,** 500–4.

Matthew, M., Hedges, R.W. and Smith, J.T. (1979). Types of β-lactamases

determined by plasmids in gram-negative bacteria. *J. Bacteriol.* **138,** 657–62.
Matthew, M. (1979). Plasmid mediated β-lactamases of gram-negative bacteria: properties and distribution. *J. Antimicrob. Chemother.* **5,** 349–58.
Mitsuhashi, S., Kawabe, H., Fuse, A. and Iyobe, A. (1975). Biochemical mechanisms of chloramphenicol resistance in *Pseudomonas aeruginosa.* In *Microbial Drug Resistance,* ed. S. Mutsuhashi and H. Hosimoto, pp. 515–23. University of Tokyo Press, Tokyo.
Oleinick, N. (1975). The erythromycins. In *Antibiotics,* vol. III, ed. J.W. Corcoran and F.E. Hahn, pp. 396–419. Springer-Verlag, Berlin.
Percheson, P.B. and Bryan, L.E. (1980). Penicillin binding components of penicillin susceptible and resistant strains of *Streptococcus pneumoniae. Antimicrob. Agents Chemother.* **18,** 390–6.
Rosenthal, K.S. and Swanson, P.E. (1977). Polymyxin and related peptide antibiotics. *Ann. Rev. Biochem.* **46,** 723–64.
Ryan, M.J. (1979). Novobiocin and Coumermycin A. In *Antibiotics,* vol. V, part 1, ed. F.E. Hahn, pp. 214–34. Academic Press, New York.
Sabath, L.D. (1979). Staphylococcal tolerance to penicillins and cephalosporins. In *Microbiology 1979,* ed. D. Schlessinger, pp. 299–303. American Society for Microbiology, Washington.
Sabath, L.D. (1977). Chemical and physical factors influencing methicillin resistance of *Staphylococcus aureus* and *Staphylococcus epidermidis. J. Antimicrob. Chemother.* **3,** (suppl. C) 47–52.
Shaw, W.V. (1977). Chloramphenicol acetyltransferase from chloramphenicol-resistant bacteria. In *Methods in Enzymology,* vol. XLIII, ed. J.H. Hash, pp. 737–55. Academic Press, New York.
Stone, D. and Smith, S.L. (1979). The aminoacid sequence of the trimethoprim resistant dihydrofolate reductase specified in *Escherichia coli* by R-plasmid R67. *J. Biol. Chem.* **254,** 10857–61.
Summers, A.O. and Silver, S. (1978). Microbiol transformations of metals. *Ann. Rev. Microbiol.* **32,** 637–72.
Sykes, R.B. and Matthew, M. (1976). The β-lactamases of gram-negative bacteria and their role in resistance to β-lactam antibiotics. *J. Antimicrob. Chemother.* **2,** 115–57.
Tait, R.C., Heyneker, H.L., Rodriguez, R.L., Bolivar, F., Covarrubias, A., Betlach, M. and Boyer, H.W. (1978). In *Microbiology 1978,* ed. D. Schlessinger, pp. 174–6. American Society for Microbiology, Washington.
Tomich, P.K., An, F.Y. and Clewell, D.B. (1980). Properties of erythromycin-inducible transposon Tn917 in *Streptococcus faecalis. J. Bacteriol.* **141,** 1366–74.
Thatcher, D.R. (1975). β-lactamase (*Bacillus cereus*). In *Methods in Enzymology,* vol. XLIII, *Antibiotics,* ed. J.H. Hash, pp. 640–52. Academic Press, New York.
Tybring, L. (1977). Special aspects of laboratory investigations with mecillinam. *J. Antimicrob. Chemother.* **3,** (suppl. B), 23–7.
Vaara, M., Vaara, T. and Sarvas, M. (1979). Decreased binding of polymyxin by polymyxin-resistant mutants of *Salmonella typhimurium J. Bacteriol.* **139,** 664–7.
Wehrli, W. and Staehelin, M. (1975). Rifamycins and other ansamycins. In *Antibiotics,* vol. III, ed. B.W. Corcoran and F.E. Hahn, pp. 252–68. Academic Press, New York.
Weisblum, B. (1971). Macrolide resistance in *Staphylococcus aureus.* In *Drug*

Action and Drug Resistance in Bacteria, vol. 1, *Macrolide Antibiotics and Lincomycin,* ed. S. Mitsuhashi, pp. 217–38. University of Tokyo Press, Tokyo.

Werner, R.G. and Daneck, K.H. (1980). Mechanisms of fusidic acid resistance in *Escherichia coli.* In *Current Chemotherapy and Infectious Disease,* ed. J.D. Nelson and C. Grassi, pp. 712–14. American Society for Microbiology, Washington.

Willsky, G.R., Bennett, R.L. and Malamy, M.H. (1973). Inorganic phosphate transport in *Escherichia coli:* Involvement of two genes which play a role in alkaline phosphate regulation. *J. Bacteriol.* **113,** 529–39.

Zighelboim, S. and Tomasz, A. (1980). Penicillin binding proteins of multiply antibiotic-resistant South African strains of *Streptococcus pneumoniae. Antimicrob. Agents Chemother.* **17,** 434–42.

4

Genetics of resistance to antimicrobial agents

Intrinsic resistance

The term, intrinsic resistance, is used to indicate natural resistance to antimicrobial agents possessed by the majority of the population of a bacterial species. Such resistance is obvious at the time of the initial introduction of an antimicrobial agent. It is the result of a property or properties of a bacterium specified most often by chromosomal genes.

The basis of many forms of intrinsic resistance is not understood. Chapter 5 provides a more detailed discussion of whole cell properties that influence antibiotic susceptibility. Only a few examples will be provided in this section.

Resistance to β-lactam antibiotics associated with a species such as cephalosporin resistance of *Enterobacter* species and penicillin resistance of *Pseudomonas aeruginosa* is due, in part, to chromosomally specified β-lactamases. *E. aerogenes* has a constitutive cephalosporinase and *P. aeruginosa,* an inducible β-lactamase. It is probable that almost all intrinsic β-lactam resistance is due in some part to β-lactamase activity. Very sensitive enzyme assays have shown β-lactamases are very wide spread among bacteria. It is possible by the use of isoelectric focusing to show that many species of bacteria have unique β-lactamases characteristic of that species.

The spectrum of resistance to β-lactams shown by a bacterial species is dependent on the effectiveness of β-lactam hydrolysis by a given enzyme. As noted elsewhere effectiveness of hydrolysis is dependent on the binding affinity of the β-lactam for the enzyme, the maximal hydrolysis rate and the total amount of enzyme available. However, another important factor contributing to intrinsic β-lactam resistance is the rate at which the drug diffuses through the cell wall. This is important mainly in gram-negative bacteria (see Chapter 5 for a discussion of the interplay of these factors).

In gram-negative bacteria the structure of the outer membrane is very important to intrinsic resistance to β-lactams and to other antibiotics. The outer membrane is a barrier to penetration to a variable extent depending particularly on the size, hydrophobicity and charge of the compound. These factors are considered in more detail in Chapter 5. The outer membranes of some gram-negative bacteria are apparently more permeable to hydrophobic antibiotics and thus, more susceptible to them than many of the Enterobacteriaceae. Thus, for example, *Neisseria gonorrhoeae* is more susceptible to erythromycin and rifampicin than most Enterobacteriaceae. Similarly *Hemophilus influenzae* is more susceptible to many antibiotics.

Many other chromosomally specified factors account for intrinsic resistance. Some anaerobic bacteria possess dihydrofolate reductase enzymes having a reduced affinity for trimethoprim. Anaerobic bacteria do not carry out oxygen-dependent electron transport and thus do not energize aminoglycoside transport into cells effectively. *Proteus* species possess an outer membrane that does not effectively bind polymyxins.

The properties of bacteria accounting for intrinsic resistance are diverse but are mainly species properties of the bacterium specified by chromosomal genes.

Mutational resistance

Numerous examples of resistance to antimicrobial agents resulting from mutation of chromosomal genes are known. In general, resistance produced by mutation unlike that associated with plasmids is by a recessive mechanism. The result of the mutation in most instances is to produce a gene product with reduced or absent affinity for the drug in question. Most often the product affected is the antibiotic target or a transport protein. Other possible effects of mutation might be to increase the amount of target or to reduce a cell's need for a particular metabolic product. In the normally haploid bacterial chromosome there is only one gene functional for any given cellular property. Alteration of that product causing reduced capability to bind an antibiotic means the organism will express some level of antibiotic resistance after allowing for the phenotypic lag needed to replace most or all of the normal gene product with the defective gene product.

A good example of resistance by mutation is that resulting in single-step high-level resistance to streptomycin and involving the *str*A gene. These mutations affect the S 12 protein of the 30S ribosomal subunit. Mutations result in one (or rarely two) amino acid substitution at one of two sites in the S 12 protein. The S 12 protein remains functional in the ribosome but does not allow the 30S subunit to bind streptomycin and the cell is resistant. The effect of a mutation of the S 12 protein is to cause a reduction in the rate of translation of mRNA although apparently less errors of translation are made. Mutations of the *str*A gene can also result in a dependency on streptomycin for growth.

Table 4.1. *Selected mutations causing altered susceptibility to several aminoglycosides*

Mutation and organism	Function	Reason for alteration of aminoglycoside susceptibility
*hem*A – *E. coli*	Synthesis of δ-aminolevulinic acid, decreased cytochromes	Reduced terminal electron transport and reduced aminoglycoside transport due, at least in part, to a reduced membrane potential portion of proton motive force. Resistant
NR70 – *E. coli*	Proton leaky, deficient in membrane Mg^{2+}, Ca^{2+} adenosine triphosphatase (ATPase) lack of coupling of electron transport and ATP synthesis (uncoupled)	Reduced proton motive force. Resistant (other mutants have shown a correlation between degree of proton leakiness and level of neomycin resistance)
*unc*A – *E. coli*	ATPase deficient, uncoupled, defect in α-subunit of F_1 component of ATPase	It is likely that electron transport is enhanced and transport of aminoglycosides is increased (as is, for example, proline). Hypersensitive

Table 4.1. *(Contd.)*

Mutation and organism	Function	Reason for alteration of aminoglycoside susceptibility
*unc*B – *E. coli*	ATPase normal but uncoupled. Defect in F_0 component	as *unc*A. Hypersensitive
*ubi*D – *E. coli*	Reduced synthesis of ubiquinone (~20% of normal) due to deficiency of 3-octaprenyl-4-hydroxy benzoate → 2-octaprenyl phenol	Reduced electron transport or reduced synthesis of aminoglycoside 'transporter'. Transport process is saturable. Resistant
*men*A, *ubi*A – *E. coli*	In the absence of hydroxy benzoate progressive decline of all respiratory quinones	As *ubi*D
men – *Bacillus subtilis*	Decreased menaquinone depending on level of shikimic acid in medium	As *ubi*D
cya – *E. coli*	Deficiency of adenyl cyclase	Reduced synthesis of several components of electron transport system. Resistant
crp – *E. coli*	Deficiency of cyclic AMP receptor protein	As cya. Resistant
*agl*A – *P. aeruginosa*	Markedly reduced cytochrome C_{552}, deficiency of nitrate reductase	Reduced terminal electron transport with a specific decrease of aminoglycoside transport. Resistant
*agl*C – *P. aeruginosa*	Deficiency of nitrite reductase, absent cytochrome d	Decreased terminal electron transport and decrease of transport of positively changed compounds. Resistant
417-T2 – *P. aeruginosa*	Increased nitrate reductase activity (7–10 fold)	Increased terminal electron transport and thus increased aminoglycoside transport (but also, for example, proline). Hypersensitive
Mutations affecting ribosomal protein or RNA		These lead in most cases to resistance to a specific aminoglycoside (e.g. streptomycin, kasugamycin) due to decreased binding by ribosomes.

Table 4.2. *Some mutations of* E. coli *chromosomal genes causing resistance or altered response to selected antimicrobial agents (see also table 2.2)*

Gene symbol	Map position (min)	Phenotype	Function
*amp*A	93	penicillin resistance	regulation of *amp*C
*amp*C	93	penicillin resistance	penicillinase structural gene
azi	2	altered azide or phenylethylalcohol susceptibility	filament formation at 42 °C
can	62	canavanine resistance	
*cml*A	18	altered susceptibility to chloramphenicol	
*cml*B	21	as *cml*A	
*cyc*A	94	resistance to D-cyclo-serine and D-serine	transport of D-alanine, D-serine and glycine
*dag*A	94	as *cyc*A	as *cyc*A
*env*A	2	increased susceptibility to several antibiotics	anomalous cell division
*fol*A	1	trimethoprim resistance	dihydrofolate reductase
*fol*B	1	trimethoprim resistance	regulatory gene
fus	72	fusidic acid	elongation factor G in protein synthesis
*glp*T	48	phosphonomycin resistance	α-glycerophosphate transport
*ksg*A	1	kasugamycin resistance	RNA methyltransferase for 16 S rRNA
*ksg*B	34	kasugamycin resistance	second step resistance to kasugamycin
*ksg*C	12	kasugamycin resistance	affects ribosomal protein S 2
*lin*B	(29)[a]	high resistance to lincomycin	
lir	(12)	increases susceptibility to lincomycin and/or erythromycin	
mac	(26)	erythromycin growth dependence	
mng	(39)	altered susceptibility to manganese	
*nal*A	48	altered susceptibility to nalidixic acid	DNA gyrase

Table 4.2. *(Contd.)*

Gene symbol	Map position (min)	Phenotype	Function
*nal*B	57	as *nal*A	
*nea*B	73	resistance to neamine	
*pho*S,*pho*T	82	resistance to arsenate	inorganic phosphate transport
*gme*C	73	tolerance to glycine, altered penicillin susceptibility	membrane defect
*gme*D	61	as *gme*C	as *gme*C
ras	(9)	sensitivity to UV and X-rays	
*ror*A	60	resistance to X-rays	
*rpl*D(*ery*A)	72	erythromycin resistance	L4 protein 50 S ribosomal subunit
*rpo*B(rif)	89	rifampicin resistance	*β*-subunit RNA
*rps*E(spcA)	72	spectinomycin resistance	S5 protein, 30 S ribosomal subunit
*rps*L(strA)	72	streptomycin resistance and dependence	S12 protein 30 S ribosomal subunit
*rps*Q(neaA)	72	neamine resistance	S17 protein, 30 S ribosomal subunit
*str*B	5	low level streptomycin resistance	
*tol*D	(23)	ampicillin resistance, tolerance to colicins E2 and E3	
*tol*E	(23)	as *tol*D	
*tol*F(*omp* F)	21	resistance to several antibiotics	structural gene for porin 1a
*ton*A	3	albomycin resistance	ferrichrome utilization
*uhp*T	81	phosphonomycin resistance	hexose phosphate transport
Selected additional mutations in *Salmonella typhimurium*			
several	–	increased susceptibility to hydrophobic antibacterial agents	deep rough chemotypes of lipopolysaccharide
*pmr*A	95	polymyxin resistance	

[a]Parentheses indicate approximate map locations.

Table 4.3. *Mutational resistance of importance in clinical medicine*

Resistance	Important bacteria	Mechanism
High-level streptomycin	*Streptococcus fecalis, Mycobacterium tuberculosis, Neisseria gonorrhoeae*	S 12 protein altered 30 S ribosomal subunit does not bind streptomycin
Spectinomycin	*N. gonorrhoeae*	S 5 protein (presumably) is altered
Multiple aminoglycoside	*Pseudomonas aeruginosa*	decreased accumulation of aminoglycosides with normal ribosomes and no inactivating enzymes
Penicillin	*N. gonorrhoeae*	(a) *pen*A gene mutated penicillin resistance only (b) *mtr*-2 gene – resistance to multiple antibiotics (impaired permeation) (c) *pen*B enhanced resistance to penicillin and tetracycline
	Streptococcus pneumoniae	decreased affinity of penicillin binding proteins for penicillin, multiple mutations
Methicillin	*Staphylococcus aureus*	unknown
Sulfonamide	*Neisseria meningitidis, N. gonorrhoeae, S. pyogenes, S. pneumoniae*	possibly increased synthesis of *p*-aminobenzoic acid or decreased sulfonamide permeability
Trimethoprim	*Enterobacteriaceae, S. aureus*	require thymine and bypass the site of action of trimethoprim (dihydrofolate reductase)
Rifampicin	*M. tuberculosis, N. meningitidis*	β-subunit of RNA polymerase does not bind the drug (*rpo*B in *E. coli*)
Nalidixic acid	*Escherichia coli,* other *Enterobacteriaceae*	resistant nal subunit of DNA gyrase. Low resistance may be due to a second separate mechanism
Novobiocin	*S. aureus*	resistant cou subunit of DNA gyrase

Resistance to a variety of other genes can result in some degree of resistance to streptomycin and often other aminoglycosides. These include many which affect the energy source needed for entry of these agents into cells. Thus mutations causing reduced terminal electron transport or proton leakiness of the cytoplasmic membrane, cause resistance. Mutants defective in adenyl cyclase or in cyclic AMP binding protein are also resistant because of reduced electron transport. (See Table 4.1 for a list of mutations causing aminoglycoside resistance.)

Table 4.2 lists many of the mutations which result in resistance for an antimicrobial agent. Additional mutations affecting penicillin targets are given in Chapter 2.

Mutational resistance of this type is not the major cause of resistance to antibiotics in clinical medicine. However, examples of such resistance have occurred among clinical isolates of bacteria (Table 4.3). On occasion mutational resistance to antibiotics has limited the use of antibiotics (for example rifampicin prophylaxis of meningococcal carriers) or has resulted in use of two drugs when some drugs are used in an attempt to eliminate resistant mutants (for example treatment of tuberculosis with streptomycin plus one or more other antibiotics).

Mutations involve the addition, deletion or substitution of one or only a few nucleotides. They probably occur mainly during DNA replication or repair or both. Such events have been termed micro-evolutionary to distinguish them from more major chromosomal changes resulting from inversion, duplication, deletion or transposition of nucleotide segments. The latter result principally from specialized recombination systems in bacteria. These include phage λ (as well as several other phage), phage Mu and transposable genetic elements. Phage Mu is capable of essentially random insertion into phage, plasmid or chromosomal DNA to produce mutations or several types of DNA rearrangements. Structures like phage λ, phage Mu and transposable elements function independently of general recombination mechanisms in the cell and do not require lengthy sequence homology with DNA segments with which they interact. These related elements comprise specialized recombination systems.

Plasmids and transposons

The most commonly occurring genetic basis of antibiotic resistance is the acquisition of additional DNA by bacterial cells resulting in a new cellular function. Such a function must be dominant over the existing antibiotic susceptible cell function in order that resistance be expressed. As detailed in Chapter 3 most plasmid and transposable DNA result in enzymes that:

(a) modify the antibiotic by the addition of a chemical group causing acetylation, phosphorylation or nucleotidylation of the drug
(b) hydrolyze β-lactam rings (β-lactamases) or organomercurials
(c) represent new enzymes with much reduced affinity for the antibiotic (dihydrofolate reductases, dihydropteroate synthetase)
(d) methylate specific adenine residues of 23S RNA
(e) reduce mercuric ion
(f) perform other less well defined functions.

Some plasmid and transposable genes can alter cell transport of tetracycline, cadmium and perhaps other agents.

Resistance plasmids

Plasmid DNA is DNA which can exist as extrachromosomal DNA. Resistance (R) plasmids carry one or more genes for antibiotic resistance. They may be either self-transmissible through conjugation or non-self-transmissible (nonconjugative). In the latter case transmission may result through mobilization of the plasmid by a co-existing self-transmissible plasmid or by transduction or transformation. Nonconjugative plasmids are found in a vast array of bacteria. They are of particular consequence among the staphylococci and several other gram-positive bacteria where they represent the only or predominant form of plasmid-mediated drug resistance.

Plasmids which are conjugative can undergo dissociation in some circumstances to produce separate plasmids carrying genes of conjugation (resistance transfer factor, RTF) and antibiotic resistance (r-determinants). The reverse sequence may also occur.

This phenomenon can occur in many types of bacteria which possess conjugative R-plasmids. It has been observed frequently, particularly among members of the *Salmonella* and *Proteus* genera. Coincidental transfer of r-determinants and a conjugative plasmid does not require that a recombinant molecule be formed between the two (or more) subunits. The genetic elements can be transferred independently in conjugation mediated by the conjugative plasmid. The r-determinant region of F-like plasmids is bounded by directly repeated copies of insertion sequence one (IS1) and is, thus, a transposable element. It can be transposed to plasmid or to a bacteriophage. Transposable DNA is discussed later in this chapter.

The fundamental property required by a plasmid is that it be able to replicate and segregate to successive generations of cells. Genes used for this function are replication (*rep*) genes. A variety of additional functions may also be present. These include: genes for incompatibility (*inc*), transfer (*tra*), fertility inhibition (*fin*), surface or entry exclusion as well as antibiotic resistance.

Incompatibility and host range. Incompatibility is expressed when two plasmids are unable to co-exist stably in the same cell. It indicates the replication systems of the two plasmids are similar. Conversely compatibility indicates that replication is not controlled by the same regulatory system. Most plasmids that are incompatible show significant DNA homology but such is not always so. Plasmids of Inc H_1, Inc H_2 and Inc H_3 groups are incompatible but lack DNA homology. Compatible plasmids of FI and FII groups show the reverse effect having extensive DNA homology mostly due to DNA in the transfer region of the plasmids.

Incompatibility particularly when combined with host range has been a useful method for classification of plasmids to carry out epidemiological studies of plasmid-borne resistance. Some plasmids either cannot enter or do not replicate in some types of bacteria. Conversely some plasmids have very wide host ranges (see Table 4.4). To date incompatibility classifications have been developed for plasmids in staphylococci, pseudomonads and for those transferable to *Escherichia coli* K12. Incompatibility groups

Table 4.4. *Incompatibility groups of plasmids found in* E. coli *K12,* Pseudomonas aeruginosa *and* Staphylococcus aureus[a]

(A) *E. coli* K12

Inc group	Host range	Representative plasmid (original host)	Phenotype[b]	Comments
B	*E. coli* *Salmonella* spp. *Shigella* spp.	R16 (*E. coli*)	Tra^+, Ap, Cm, Sm, Su, Tc	Mol. wt 69×10^6; G + C 50%; *bla* = OXA-2; B pilli
C	*E. coli, Providencia* sp., *S. typhimurium, P. mirabilis, V. cholerae, P. vulgaris, K. pneumoniae, S. marcescens*	R40a (*S. typhimurium*)	Tra^+, Ap, Km, Pm, Su	Mol. wt 96×10^6; *bla* = TEM-1; aphA; includes plasmids previously incE; includes RA1 – previously incA; C pili
D	*E. coli, Providencia* sp.	R711b (*Provindencia* sp.)	Tra^+, Km	
FI	*E. coli, P. morganii, P. vulgaris, S. typhimurium, Rhizobium lupini, Providencia* sp., *Shigella* spp., *Enterobacter cloacae, P. mirabilis*	R386 (*E. coli*)	Tra^+, Tc, Phi, (T7), Dps, (f1, f2)	Derepressed pilus synthesis ($fin0^-$)
		F (*E. coli* K12)	Tra^+, Ph, (T3T7 II), Dps, (F1, F2, F17, MS2)	Mol. wt 63×10^6; G + C 49%; depressed pilus synthesis ($fin0^-$); F pili
F II	*E. coli, S. typhimurium, S. flexneri, K. pneumoniae, P. mirabilis, P. vulgaris, P. morganii, Providencia* sp., *S. typhi, S. marcescens, S. paratyphi*	R1 (*S. paratyphi* B)	Tra^+, Ap, Cm, Km, Sm, Sp, Su, Dps, (f1, f2, MS2, M13)	Mol. wt 62×10^6; G + C 52%; F pili
F III	*E. coli*	ColB-k98 (*E. coli*)	Dps (f1, f2), Cba	Mol. wt 70×10^6

F IV	*S. typhimurium*	R124 (*S. typhimurium*)	Tra^+, Tc, Dps (f1, F2, M13, MS2)	
H 1	*S. typhi* *E. coli*	TP123 (*S. typhi*)	Tra^+, Sm, Cm, Su	Little or no DNA homology with H-2 or H-3; may displace F factor
H 2	*S. typhi* *E. coli*	TP116 (*S. typhi*)	Tra^+, Sm, Cm, Su	Little or no DNA homology with H-1 or H-3; thermosensitive transfer
H 3	*S. ohio* *E. coli*	MIP233 (*S. ohio*)	Lac, Sm, Cm, Su	Mol. wt 143.7×10^6; pili specified by H plasmid; little or no DNA homology with H-1 or H-2
I α	*E. coli, Salmonella, Shigella, Klebsiella*	R64 (*S. typhimurium*)	Tra^+, Sm, Tc, Phi, (τ), Dps (If1)	Mol. wt 72×10^6; G + C 50%; *aph*C; I pili
I2	*E. coli*	TP114 (*E. coli*)	Tra^+, Km, Dps (If1)	Mol. wt 41×10^6
I γ	*E. coli, S. typhimurium*	R621a (*S. typhimurium*)	Tra^+, Tc, Dps (If1)	Mol. wt 65×10^6; I pili
I δ	*E. coli, S. typhimurium*	R821a (*S. typhimurium*)	Tra^+, Ap, Dps (If1)	Mol. wt 43×10^6
I ς	*E. coli, S. typhi*	R805a (*S. typhi*)	Tra^+,Km	Mol. wt 48×10^6
J	*Proteus* spp., *E. coli*	R391 (*P. rettgeri*)	Tra^+, Km, Nm, Hg	*aph*A; I pili – identified
K	*S. flexneri, E. coli*	R387 (*S. flexneri*)	Tra^+, Cm, Sm	*cat;* K pili identified
M	*S. paratyphi, E. coli, P. morganii, S. marcescens, K. pneumoniae*	R69 (R1P69) (*S. paratyphi* B)	Tra^+, Ap, Km, Nm, Pm, Tc	bla = TEM-1; *aph*A; includes plasmids previously incL; M pili identified

Table 4.4. *(Contd.)*

(A) *E. coli* K12 Inc group	Host range	Representative plasmid (original host)	Phenotype[b]	Comments
N	*Proteus, Klebsiella* spp., *Salmonella* spp., *Providencia* sp., *Shigella* spp., *E. coli*	N-3 (*Shigella*)	Tra^+, Sm, Sp, Su, Tc, Hg, Dps (Ike)	Mol. wt 33×10^6; G + C 50%; *aad*A; large numbers of plasmids are in the group
P	Very wide: *E. coli, Pseudomonas* spp., *Serratia* spp., *Bordetella, Bronchiseptica, Proteus* spp., *Rhizobium* spp., *Salmonella* spp., *Shigella* spp., *Providencia* sp., *Neisseria* spp., *Rhodopseudomonas* spp., *Azatobacter, Caulobacter, Klebsiella* spp., *Agrobacterium tumefaciens, Chromobacterium violaceum*	RP4 (*P. aeruginosa*)	Tra^+, AP, Km, Nm, Tc, Dps (PRD1, PRP1, PR3 PR4, PR5)	Mol. wt 36×10^6; G + C 58%; *bla* = TEM-2; *aph*A; P pili
Q	*E. coli, Salmonella* spp., *Proteus* spp., *Providencia* sp., *Pseudomonas* spp.	R1162 (*P. aeruginosa*)	Tra^-, Sm, Su	Mol. wt 5.5×10^6
T	*E. coli, Proteus* spp., *Providence, K. aerogenes, S. typhimurium*	Rts1 (*P. vulgaris*)	Rep (ts), Tra^+, Km	G + C 45%; *aph*A; T pili

V	*P. mirabilis, P. vulgaris, E. coli*	R753 (*P. mirabilis*)	Tra^+, Ap, Cm, Sm, Su	Mol. wt 68×10^6; *bla* = OXA-I; V pili identified
W	*E. coli, S. flexneri, Proteus* spp., *Klebsiella* spp., *P. aeruginosa, Salmonella* spp., *A. liquefaciens*	S-a (*S. flexneri*)	Tra^+, Cm, Km, Sm, Su, Dps (PRD1)	Mol. wt 23×10^6; G + C 62%
X	*P. morganii, E. coli, Providencia* sp.	R6K (*P. rettgeri*)	Tra^+, Ap, Sm	Mol. wt 26×10^6; G + C 45%; *bla* = TEM-1; X-pili detected
(B) *Pseudomonas aeruginosa* and other Pseudomonads				
P-1	See *E. coli* P	RP4 (*P. aeruginosa*)	See *E. coli* P	IncP-1 of *Pseudomonas* is identical to P of *E. coli*
P-2	*Pseudomonas* spp., *Flavobacter meningoosepticum*. Not *E. coli*	R931 (pLB931) (*P. aeruginosa*)	Tra^+, Sm, Tc, Hg, Phi (B3, B39, E79, F116L, G101.M6pB1)	Have very large mol. wt; G + C 59%; *aph*C; morphological detection of pili; very large plasmid group; members often specify tellurium resistance
P-3	*Pseudomonas* spp., *E. coli* (see *E. coli* C)	R40a (*P. aeruginosa*)	See *E. coli* C	IncP-3 = IncC of *E. coli*
P-4	See IncQ of *E. coli*	R1162 (*P. aeruginosa*)	See *E. coli* Q	Inc P-4 = Inc Q of *E, coli;* nonconjugative
P-5	*P. aeruginosa*	Rms 163 (*P. aeruginosa*)	Tra^+, Cm, Su, Tc	
P-6	*P. aeruginosa*	Rms 149 (*P. aeruginosa*)	Tra^+, Cb, Gm, Sm, Sp, Su, Phi (F116)	*bla; aac*C; no Sm inactivating enzyme found
P-7	*P. aeruginosa*	Rms 148 (*P. aeruginosa*)	Tra^+, Sm, Phi (B39, C5)	*aph*C

Table 4.4. *(Contd.)*

(B) *Pseudomonas aeruginosa* and other Pseudomonads		Representative plasmid (original host)	Phenotype[b]	Comments
P-8	*P. aeruginosa*	RP2 (*P. aeruginosa*)	Tra^+, Hg	Mol. wt 59×10^6; G + C 58%; mobilizes chromosome of some *P. aeruginosa*
P-9	*P. aeruginosa*	pMG38 (*P. aeruginosa*)	Tra^+, Gm, Km, Su, Tc, Tm, Hg	includes SAL plasmid
P-10	*P. aeruginosa*	pMG41 (*P. aeruginosa*)	Tra^+, Gm, Km, Sm, Su, Tm, HG	Mol. wt 63×10^6
P-11	*P. aeruginosa*	pMG39 (*P. aeruginosa*)	Tra^+, Cb, Gm, Km, Sm, Tm	Mol. wt 57×10^6
(C) *Staphylococcus aureus*				
1	*S. aureus*	pI258 (*S. aureus*)	Tra^-, Pc, Asa, Asi, G, Cd, Pb, Bi, Em	Mol. wt 18.4×10^6; G + C ~ 35%
2	*S. aureus*	pII147 (*S. aureus*)	Tra^-, Pc, Asa, Hg, Cd, Pb	Mol. wt 21×10^6; G + C 35%
3	*S. aureus*	pT127 (*S. aureus*)	Tra^-, Tc	Mol. wt 2.7×10^6
4	*S. aureus*	pC221 (*S. aureus*)	Tra^-, Cm	Mol. wt 30×10^6; G + C 35%
5	*S. aureus*	pS177 (*S. aureus*)	Tra^-, Sm	Mol. wt 2.7×10^6; G + C 35%
6	*S. aureus*	pK545 (*S. aureus*)	Tra^-, Km, Nm	Mol. wt 15×10^6
7	*S. aureus*	pUB101 (*S. aureus*)	Tra^-, Pc, Cd, Fa	Mol. wt 14.6×10^6

[a]Many plasmids cannot be classified into the groups given. Several plasmids have been detected in gram positive bacteria other than *S. aureus*.

[b]Abbreviations and symbols. Antibiotic and metal symbols refer to resistance to those agents.

*aac*C, aminoglycoside 3-*N*-acetyltransferase; *aad*A, aminoglycoside 3″-adenylyltransferase; Ap, ampicillin (β-lactamase); *aph*A, aminoglycoside 3′-phosphotransferase; *aph*C, aminoglycoside 3″-phosphotransferase; Asa, arsenate; Asi, arsenite; Bi, bismuth ion; *bla*, β-lactamase; *cat*, chloramphenicol acetyltransferase; Cb, carbenicillin (β-lactamase); *cba*, production of colicin B; Cd, cadmium; Cm, chloramphenicol; Dps, donor phage sensitivity; Em, erythromycin; Fa, fusidic acid; Gm, gentamicin; Hg, mercuric ion; Km, kanamycin; Lac, lactose fermentation; Nm, neomycin; Pc, penicillin (β-lactamase); Phi, interference with phage production; Pm, paromonycin; Pma, resistance to phenyl mercuric acetate; Rep(ts), replication thermosensitive; Sm, streptomycin; Su, sulfonamide; Tc, tetracycline; Tm, tobramycin; Tra, transfer.

of these three groups of organisms are given in Table 4.4 as are some of their properties.

Compatibility of two plasmids should be confirmed by the separate transfer of each plasmid out of the cell as recombination of plasmids can occur and can mimic stable co-existence of antibiotic resistance markers. Recombination is most likely to occur if selection has been made for resistance markers of both plasmids in the recipient cell. Other difficulties seen in defining incompatibility groups include: transposition of DNA (see later in this chapter); strong surface exclusion preventing entry of a second but compatible plasmid; cryptic plasmids transferred from wild type strains with the plasmids under study and influencing compatibility with other plasmids; initial dislodgement of a compatible plasmid upon entry of the second plasmid; variations in degree of incompatibility and thus, the rate and extent of loss of a second plasmid; asymmetrical loss of plasmids with one plasmid being 'superior' and preferentially retained; recombined plasmids with more than a single compatibility function.

Plasmids of many incompatibility groups have been shown to specify the production of pili. It is probable that all conjugative plasmids specify pili. In general in those cases examined to date, pili of different incompatibility groups are different. F pili are determined by members of the F group (FI to FIV). Pili are also determined by plasmids of the I complex, Inc B, Inc C, Inc D, Inc H, Inc H_2, Inc J, Inc K, Inc M, Inc N, Inc P, Inc T, Inc V, Inc W and Inc X. Pili are determined by the *Pseudomonas* group P-2 (as well as P-1 which is the same as P).

The sides of F and P pili serve as receptors for RNA phages like MS2 and PRR1 respectively. Phage 1F1 attaches to tips of I pili and the filamentous DNA phage fd attaches to tips of F and Inc D pili. The lipid containing DNA phage group PR3, PR4 and PR5 attach to the tips of P, N and W pili. Thus pili tips serve as less specific phage receptors than do pili sides. Phage IKe are filamentous DNA phages specific for N group plasmids.

Plasmids for the most part also show surface exclusion inhibiting entry of incompatible plasmids of the same Inc group. However, surface exclusion is not always demonstrable and can be directed towards compatible but related plasmids (e.g. F and I group plasmids).

This is no correlation between Inc groups and pattern of antibiotic resistance carried.

General molecular properties. The size of plasmids carrying resistance determinations varies greatly. It depends upon whether the plasmid also contains the information required for conjugation, the so called RTF components. Conjugative R plasmids are generally above 25×10^6 molecular weight and some reach about 300×10^6 molecular weight. Molecular weights of various plasmids are given in Table 4.4. Nonconjugative resistance plasmids are much smaller and usually range from 1×10^6 to 10×10^6 molecular weight although larger weights are observed for example with some plasmids of *S. aureus*.

The number of copies of R-plasmid per cell is related to the molecular weight. Larger plasmids tend to have a stringently controlled copy number with 1 or 2 copies of the plasmid as covalently closed circular (CCC) DNA per cell. Small plasmids, normally nonconjugative, are present with from 5 to 30 copies per cell. R6K, a conjugative plasmid of the X Inc group, has a molecular weight of 26×10^6 but has a copy number of 13 to 38. This is an exception for a conjugative plasmid and copies of this order are usually seen with small conjugative plasmids.

Increased levels of resistance to some antibiotics particularly chloramphenicol and streptomycin have been achieved by growth of R^+ bacteria in the presence of one of these drugs or by selection of mutants with heightened resistance. For chloramphenicol this is clearly associated with elevated levels of chloramphenicol acetyl transferase. The mechanism of such heightened resistance varies for the antibiotic and among plasmids and particularly bacteria. In *P. mirabilis* heightened resistance has been shown to be due to multiple copies of either the whole r-determinant (the part of the R plasmid carrying resistance genes) or a specific resistance gene in association with the RTF. This forms enlarged R-plasmids. In *E. coli* it appears due to multiple copies of the whole R-plasmid or to copies of small pieces of DNA. In the last case these apparently specified heightened ampicillin resistance.

Plasmid DNA can be isolated from cells in the covalently closed circular (CCC) or supercoiled form. The CCC form allows the isolation of plasmid DNA by methods involving reversible denaturation due to the failure of CCC DNA to unwind and separate completely whereas open circular (OC) (one strand nicked) or linear plasmid or chromosomal DNA fail to renature. Ethidium bromide binding to CCC DNA is restricted relative to OC or linear DNA. Thus the buoyant density of CCC DNA is reduced less than that of OC or linear DNA in ethidium bromide cesium chloride density gradient centrifugation. CCC DNA has a higher buoyant density in alkaline cesium chloride density gradients. These and other principles are useful to separate CCC DNA from linear chromosomal DNA broken during extraction and from OC and linear plasmid DNA.

Most plasmid CCC DNA can be converted to the open circular form through treatment with detergents or proteolytic enzymes. CCC DNA exists, at least in part, in a DNA–protein complex which when treated with the preceding agents causes relaxation of the CCC form to the OC form.

Replication of plasmid DNA is a complex subject and beyond the scope of this chapter. However, a few general statements can be made. Replication of all plasmids is not identical and can vary in several respects.

Conjugative plasmids like R6K and a derivative of R6K, RSF 1040, have two origins of replication and are replicated bidirectionally from the origin. For nonconjugative plasmids like Col E1 there is a single origin and unidirectional replication whereas RSF 1010 has a single origin but bidirectional replication.

Inhibition of protein synthesis by chloramphenicol allows completion of a current round of chromosomal or large molecular weight plasmid DNA replication but does not allow initiation of new round of replication. Many small nonconjugative plasmids can initiate new replication under these conditions and can continue replication for sometime. Col E1 can replicate for several hours and produce close to 1000 copies per chromosome.

RNA synthesis has also been shown essential for the replication of many plasmids including R6K, RSF1030, F and Col E1. This seems necessary for the synthesis of an RNA primer needed to allow subsequent DNA synthesis. The primer is apparently covalently incorporated into the daughter plasmid and its excision requires ongoing protein synthesis.

DNA polymerase III is necessary for chromosomal and most plasmid replication whereas polymerase I is not required. However, Col E1 and RSF 1030 require polymerase I but not polymerase III (these two plasmids also replicate freely in the presence of chloramphenicol). RSF 1010 is somewhat unstable in polymerase I mutants.

Some naturally occurring plasmids are thermosensitive for replication. An example is R-plasmid Rts1 from *P. mirabilis*. Cells harboring Rts1 have normal growth at 32 to 37 °C but at 42 °C there is no increase or only a linear increase in viable cell counts. H plasmids are thermosensitive for transfer in that they transfer poorly at 33 °C but efficiently at 25 °C.

R-plasmids may be lost from cells spontaneously upon storage or may be unstable at elevated temperatures. They may be artificially eliminated by selective inhibition of plasmid replication by curing agents. Curing conditions include DNA-intercalating compounds such as ethidium bromide and acriflavine, inhibition of RNA synthesis by rifampicin and thymine deprivation. Cells containing R-plasmids can be selectively killed by agents like EDTA, macarbomycin or sodium dodecylsulfate. Macarbomycin kills bacteria producing pili.

Genetics of R-plasmids. Most of the genetic information carried by conjugative plasmids relates to the plasmids's transfer system. The most intensive study of transfer genes has been done for F factor and is the basis of the description given here. A map of R-factor R 100 is provided in Fig. 4.1. Four transfer systems have received some degree of study. These include besides F, I, P and N. Other transfer systems also exist.

An F-type transfer system with a mutation in one of *tra* (for transfer) A, *tra*B, *tra*C, *tra*E, *tra*F, *tra*H, *tra*K, *tra*L and some in *tra*C genes does not produce pili. The structural gene for the subunit protein of pili, pilin is *tra*A. The function of the other genes involves pilin modification and assembly but their exact role is unknown.

The function of other gene products is known to a greater or lesser extent and is given in Table 4.5. The other transfer systems are probably genetically unrelated to that of F. The transfer system of the P-plasmid RP1 contains two separated transfer regions each comprising several genes. The F transfer system is organized as an operon. The *tra*J gene product is required for expression of the transfer operon.

Plasmid transfer inhibition occurs for most plasmids after initial plasmid entry into a cell and a short period of cell growth. In the F system this is due to the *fin* (fertility inhibition) O and P genes which form the FinOP system. This system prevents expression of the *tra*J gene product and, thus, the transfer operon. The F-plasmid is naturally *fin*O^{-} and does not inhibit its own transfer. Systems subject to fertility inhibition are often referred to as being repressed. Mutants which are apparently *fin*O^{-} are spoken of as being derepressed. Plasmids termed fi^{+} in older literature are those inhibiting transfer of F and are part of the F complex whereas fi^{-} plasmids do not inhibit F transfer. They represent other transfer systems (e.g. P and I). However, fi is used in a more general sense to mean fertility inhibition of a specific transfer system. Thus a plasmid could be fi^{+}, for RP4 specified transfer. This means the plasmid inhibits RP4 specified fertility (transfer).

Plasmid transfer is much higher (high-frequency transfer, HFT) in the time immediately following plasmid entry into a cell before

Fig. 4.1. Genetic map of R-factor R100. For symbols see tables 4.5. and 4.6; op, operator; expanded portion of map is the mer operon. Reproduced, with permission, from the *Annual Review of Microbiology*, vol. 32. © 1978 by Annual Reviews Inc.

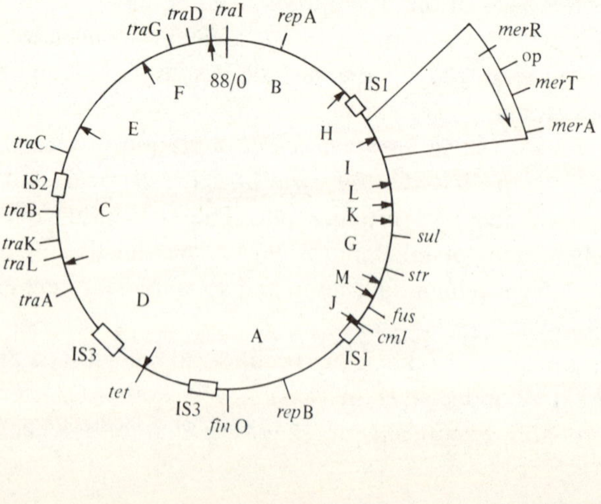

Table 4.5. *Function of genes carried by conjugative R-plasmids*

Genotype	Function
(A) transfer (tra)[a]	
*tra*A	pilin synthesis
*tra*B, C, E, F, H, K, L, part of *tra*G	pilin modification and pilus assembly
*tra*I, *tra*D, *tra*G (part), *tra*M	DNA transfer genes
*tra*S, *tra*T	surface exclusion
*tra*J	positive control gene
(B) transfer control	
*fin*O	inhibits transfer of F type plasmids
*fin*P	inhibits transfer of specific F type plasmids. *Fin*O and *fin*P form the FinOP system with prevents expression of *tra*J
(C) Nucleic acid functions	
inc	incompatibility
rep	replication
phi	phage inhibition
ori	origin of transfer replication
hms	modification of DNA by methylation
hsr	restriction of DNA by endonuclease activity
(D) Antibiotic resistance	
aph	aminoglycoside phosphotransferase
acc	aminoglycoside acetyltransferase
aad	aminoglycoside adenylyltransferase
ant	aminoglycoside nucleotidyltransferase
bla	β-lactamase
cat	chloramphenicol acetyltransferase
asa	arsenate resistance
asi	arsenite resistance
bis	bismuth ion resistance
cad	cadmium ion resistance
cob	cobaltous ion resistance
erm	erythromycin resistance
fus	fusidic acid resistance
mer	mercuric ion resistance
nic	nickelous ion resistance
ant	antimony ion resistance
sul	sulfonamide resistance
tet	tetracycline resistance
dfr	dihydrofolate reductase
(E) Other miscellaneous phenotypes may also be expressed	

[a]Transfer genes are those described for an F factor.

transfer inhibition is expressed. This seems due to transient synthesis of the *tra*J gene product and before it is extensively diluted by cell growth. After some hours residence in a cell the frequency at which a plasmid is transferred in bacterial matings falls substantially. The N and P transfer systems do not apparently have transfer inhibition systems.

R-plasmids have been shown to specify restriction-modification enzymes. Restriction enzymes are endonucleases which recognize specific nucleotide sequences and generally cleave DNA at a specific site. Modification enzymes normally recognize the same nucleotide sequence and modify a specific nucleotide. This process is a mechanism by which DNA is identified and its degradation as foreign DNA is prevented. Restriction endonucleases have obtained great utility in physical DNA mapping and in preparation of recombinant DNA.

R-factors can promote transfer of the chromosome. Transfer may be polarized and associated with recombination between plasmid and chromosome. It may also be non-polarized not involving recombination and generally occurring at low frequency. In the case of F, recombination with the chromosome may occur with increased frequency at certain locations ('hotspots'). Identical insertion sequences in the chromosome and plasmid act as recombinational hotspots with the F factor forming a high-frequency transfer system (HFT).

Transposable elements

Transposable elements are discrete pieces of DNA capable of serial translocation from one replicon to another. They have characteristic structures and preserve their structure on translocation. Translocation does not require the *rec*A specified recombination system. They specify self-encoded phenotypic traits including antibiotic resistance, resistance to heavy metals, enzymes involved in metabolism of lactose, raffinose, toluene, xylene or salicylate, enteroxin production, K88 bacterial surface antigen, histidine genes of *E. coli* K12, inducible transposition and probably many other phenotypes. Conceivably any gene of a chromosome, plasmid or virus replicon could be located on a transposon. Transposons may produce mutations, DNA rearrangements (e.g. deletions, inversions) and some can behave as transcriptional start and stop signals.

Transposition may be to random sites or at preferred sites in the recipient replicon. This varies depending on the transposable element.

A hierarchy of transposable elements exists. The simplest are insertion sequences (IS) elements which do not encode known phenotypic traits (see Table 4.6). The second group are larger transposons some of which contain IS units at their termini (Table 4.7). The third is represented by phage Mu which can act both as a transposon and a cell-free virus. Table 4.7 lists many properties of transposable elements.

Transposons have characteristic structures. Almost all have inverted repeat DNA sequences at their termini. Apparent exceptions to this include transposon (Tn) 9, Tn(Raf) and Tn(R-det) which contain direct repeats of IS1 at their termini. However, IS1 contains inverted repeat terminal sequences, meaning that all IS and transposable elements have terminal repeat sequences.

Table 4.6. *Properties of transposable elements – IS elements*

IS element	Size (base pairs)	Terminal repeated sequences	Length of recipient DNA duplication	Recipient genomes carrying element
IS1	768	28 of first 34, I[a]	9	*E. coli, S. typhimurium, Citrobacter,* several phage and plasmids
IS2	1327	32 of first 41, I	5	*E. coli,* phage λ, plasmid F, R6, R100
IS3	1200	not known	not known	*E. coli,* F plasmid
IS4	1400	not known	11	*E. coli*
IS5	1250	50, I	not known	*E. coli,* phage
	5800	35, I	5	*E. coli,* F plasmid
ISR1	1100	not known	not known	*Rhizobium lupini*
ISP1	1800	not known	not known	*Pseudomonas aeruginosa*

[a]I = inverted.

Table 4.7. *Properties of transposable elements*

Transposon	Size (kilo base pairs)	Terminal repeated sequences[a]	Markers[b]	Original plasmid source	Length of recipient DNA duplication	Transposon frequency	Comments[c]
Tnl	4.6	~40, short, I[a]	Ap	RP4-*Pseudomonas*	5	10^{-2}	TnA group, TEM β-lactamase, cleaved BamHI, HaeII, III, HincII
Tn2	4.6	~40, short, I	Ap	RSF1030-*Salmonella*	5		In A group
Tn3	4.6	38, short, I	Ap	R1-19-*Salmonella*	5	10^{-2} to 10^{-5}	In A group
Tn401,801,802	4.6	~40, short, I	Ap	RP1-*Pseudomonas*	5	10^{-2} to 10^{-4}	In A group
Tn901,902	4.6	~40, short, I	Ap	pR130; *Salmonella* -phage P7	5?		In A group
Tn1701	4.6	~40, short, I	Ap	NTP1-*E. coli*	5?		In A group
Tn4	20.5	40?, short, I	Ap, Sm, Su, Hg	R1-19-*Salmonella*		10^{-6} to 10^{-7}	Cleaved by BamHI, HaeII, III, HincII
Tn21	15.7	<140, short, I	Sm, Su	R100-1-*Salmonella*			Cleaved by EcoR1
Tn(AB)	14.7	—	Ap, Su	R938-*Serratia*			
Tn5	5.2	1450, long, I	Km	JR67-*Klebsiella*	9	10^{-2} to 10^{-3}	Cleaved by HindII, III
Tn6	5.1	short	Km	JR72-*E. coli*			
Tn601,903	3.1	1000, long, I	Km/Nm	R6; R6-5-*Salmonella*	9		Cleaved by BpaII, HaeIII, HapII, HgaI, HhaI, HindIII
Tn904	5.6	—	Sm	pMG-1-*Pseudomonas*		—	
Tn7,71,72	12.7	short, I	Tp, Sm	R483, R721, pBW1-*E. coli*		5×10^{-4}	Cleaved by BamHI, EcoR1, HindIII

Tn402	7.5	none detected	Tp	R751-*Klebsiella*			
Tn9	2.5	IS1, long, D	Cm	R10-*Shigella*	9		Cleaved by BalI, EcoR1
Tn168	2.1	IS1, long, I	ST toxin	ST-*E. coli*	9		Cleaved by HincII, ST – heat stable enterotoxin of *E. coli*
Tn(R-det)	23	IS1, long, D	Cm, Sm, Su	R100-1-*Shigella*	9?		
Tn(Raf)	40	IS1, long, D	See comments	pRSD2-*E. coli*			Markers-raffinose fermentation, H_2S formation, K88 antigen production
Tn10	9.3	1.4, long, I	Tc	R100-*Shigella*	9	10^{-6} to 10^{-7}	Cleaved by many restriction endonucleases
Tn1721,1771	10.8	< 50, short, I	Tc	pRSD1-pS202-*E. coli*			Likely identical transposons, cleaved by EcoR1, Sma1, HindIII
Tn551	5.2	< 100, short, I	Em	pI258-*Staphylococcus*		10^{-4} to 10^{-5}	
Tn917	4.5	280, I	Em	PAD2-*Streptococcus fecalis*			Inducible for increased frequency of transposition
Tn501	5.2	< 150, short, I	Hg	pVS1-*Pseudomonas*		10^{-1} to 10^{-2}	Cleaved by EcoR1, HindIII, SalI, PstI, SacII, SmaI
Tn951	16.6	100, short, I	Lac	PGC1-*Yersinia*			Cleaved by BamHI, EcoR1, HindIII, PstI, lactose fermentation
Tn(Tol)	52.5	—	Xyl, Tol	Tol-*Pseudomonas*			Xylene and toluene catabolism
Tn(Sal)	30	—		Sal-*Pseudomonas*			Salicylate degradation
Tn(his-gnd)	44	1400, long, I		*E. coli*			Histidine genes of *E. coli*
Tn(Ti)	16.5		Tumor induction	Ti plasmid agrobacterium			Tentatively included as a Tn

[a]I, inverted; D, direct.

[b]Ap, ampicillin (β-lactamase); Sm, streptomycin; Su, sulfonamide; Hg, mercury; Km, kanamycin; Nm, neumycin; Tp, trimethoprim; Cm, chloramphenicol; Tc, tetracycline; Em, erythromycin; Lac, lactose fermentation; Xyl, xylene catabolism; Tol, toluene catabolism; ST, heat stable toxin.

[c]Known cleavages by restriction endonucleases are provided.

Fig. 4.2 is a diagram of the general structure of a transposable DNA element. The phenotypic traits if present are encoded by the central DNA sequences. This includes, for example, a gene for *β*-lactamase production. However, this portion of the transposon also probably specifies functions essential to transposition. Tn3 is a small transposon with 38 base pair inverted repeats at its termini. The central sequence specifies a TEM *β*-lactamase in addition to other polypeptides. One is a 19 000 molecular weight polypeptide which is a regulatory protein. A site adjacent to that coding for the regulatory protein is important for fidelity of transposition. In addition, a second large polypeptice has been described which is

Fig. 4.2. Transposition process and the general structure of a transposable element. (*a*) Recipient DNA cleavage as an initial step in transposition. (*b*) Double-stranded recipient replicon containing a transposon showing the general structure of a transposable element. (*c*) Single-stranded transposable element with double-stranded stem (hairpin or stem-loop) produced following denaturation and intrastrand annealing of the structure shown in (*b*). The hairpin loop in readily identified by electron microscopy.

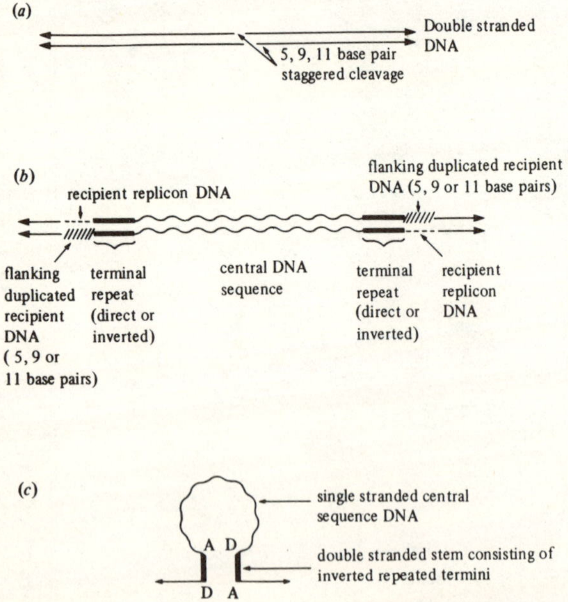

about 105 000 molecular weight. Impairment of its function leads to a transposition-deficient phenotype.

Studies of deletions and insertions of Tn10 have shown that deletions affecting the central, nonrepeated sequence did not affect the frequency of transposition. These studies have suggested that one of the repeated insertion sequences encodes an essential transposition function which is expressed less efficiently by the opposite insertion sequence.

The repeated sequences flanking transposable elements fall into two classes based on the length of their repeated sequences. Group 1 has short sequences of 50 base pairs or less. Examples are IS1 and Tn2. The second has longer repeats greater than 700 base pairs. This groups includes Tn5, Tn9 and Tn10. It is likely the shorter repeats provide recognition signs for site-specific recombination. The longer repeats appear to do this and also encode functions involved in translocation.

In the process of transposition (Fig. 4.2) transposons generate a 5, 9, or 11 base pair duplication on insertion of the transposon. This apparently results from two staggered enzymic single-stranded cleavages overlapping by 5, 9 or 11 base pairs depending on the specific transposon. The complementary DNA is newly synthesized producing a repeated sequence joined directly to the inserted element. On excision one of them is removed with the transposable element (see Fig. 4.2).

Upon movement of a transposable element to a new site the inserted transposon is not lost from the original site. Thus replication and transposition are linked processes. Models for transposition of DNA have been proposed by Grindley and Sherratt and by Shapiro. The interested reader is referred to those publications.

Plasmid diversity is in part related to transposable elements. A study of three R-plasmids isolated from different parts of the world has shown the role that transposons have played in the evolution of antibiotic resistance plasmids. R1 from England, R6 from Germany and R100 (=NR1, R222) from Japan are FII R-plasmids which carry the same transposons. The structures of portions of these plasmids are shown in Fig. 4.3. All three contain the same chloramphenicol transposon and streptomycin (Sm)–

sulfonamide (Su) transposon. However in R1, Tn3 is inserted in the Sm–Su transposon and specifies β-lactamase (ampicillin (Ap) resistance). This combination of Sm, Su and Ap is Tn4. R1 also has a segment coding neomycin (Nm)/kanamycin (Km) resistance which is not known to be a transposon. Tn903 specifying Nm/Km resistance is inserted into R6 as shown in Fig. 4.3. As noted the whole set of r-determinants is transposable. Finally, the RTF components of R6 and R100 carry Tn10 specifying tetracycline resistance.

The combination of plasmids which are transferable between bacteria and transposable drug resistance is one that can explain the rapid dissemination of antibiotic resistance among many types of bacteria. As noted earlier some plasmids have very wide host ranges and can be conjugated to many different bacterial species. P-plasmids are a particularly impressive example. At this time transposons have been detected in *E. coli, Klebsiella, Salmonella, Shigella, Proteus, Citrobacter, Serratia, Yersinia, Hemophilus, Pseudomonas, Rhizobium, Staphylococci* and *Streptococci.* They probably occur in all types of bacteria. It is therefore not surprising

Fig. 4.3. Transposable elements in R-plasmids, R1, R6 and R100. Cm, chloramphenicol; Sm, streptomycin; Su, Sulfonamide; AP, ampicillin; Km-Nm, kanamycin-neomycin; Tc, tetracycline; RTF, resistance transfer factor; Tn, transposon.

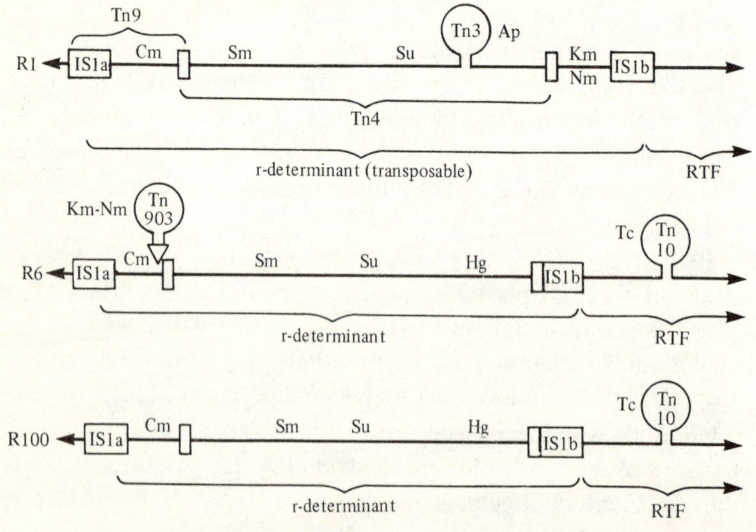

that resistance to antimicrobial agents is widely disseminated and that many identical mechanisms are found in widely different bacteria.

Transposons specifying ampicillin resistance include Tn1, 2, 3, 401, 801, 802, 901, 902 and 1701. These appear to be homologous elements expressing ampicillin resistance and are collectively termed TnA. Interestingly it has been shown that some plasmids carry only fragments of TnA. Plasmid PVE445 from a *Hemophilus influenzae* strain isolated in West Germany for example has only 30–40% of TnA. A plasmid isolated from *Hemophilus ducreyi* in Canada however has 100% of TnA. Part TnA structures like that of PVE445 cannot transpose although they could form a part of a larger transposable unit. Some transposons are composites of two transposons. For example Tn4 consists of Tn3 inserted into Tn21. Table 4.7 provides other properties of some transposons and Table 4.6 properties of insertion sequences.

Selected references

Achtman, M., Skurray, R.A., Thompson, R., Helmuth, R., Hall, S., Beutin, L. and Clark, A.J. (1978). Assignment of *tra* cistrons to *Eco* R_1 fragments of F sex factor DNA. *J. Bacteriol.* **133,** 1383–92.

Bachmann, B.J. and Low, K.B. (1980). Linkage map of *Escherichia coli* K12, edition 6. *Microbiol. Rev.* **44,** 1–56.

Barth, P.T., Grinter, N.J. and Bradley, D.E. (1978). Conjugal transfer system of plasmid RP4: analysis by transposon 7 insertion. *J. Bacteriol.* **133,** 43–52.

Bradley, D.E (1980). Determination of pili by conjugative bacterial drug resistance plasmids of incompatibility groups B,C,H,J,K,M,V and X. *J. Bacteriol.* **141,** 828–37.

Bradley, D.E., Raizens, E., Fives-Taylor, P. and Ou, J. (Eds.) (1978). *Pili. International Conferences on Pili.* Washington, DC.

Bryan, L.E. (1980). Aminoglycoside-resistant mutation of *Pseudomonas aeruginosa* defective in cytochrome C_{552} and nitrate reductase. *Antimicrob. Agents Chemother.* **17,** 71–9.

Bukhari, A.I., Shapiro, J.A. and Adhya, S.L. (Ed.) (1977). *DNA Insertion Elements, Plasmids and Episomes.* Cold Spring Harbor Laboratory, New York.

Crosa, J.H., Luttropp, L.K. and Falkow, S. (1978). Use of autonomously replicating nine R6K plasmids in the analysis of the replication regions of the R plasmid R6K. *Cold Spring Harbor Symp. Quant. Biol.* **43,** 111–20.

Datta, Naomi. (1979). Plasmid classification: Incompatibility grouping. In *Plasmids of Medical Environmental and Commercial Importance,* ed. K.N. Timmis and A. Pühler, pp. 3–12. Elsevier/North Holland Biomedical Press.

Falkow, S. (1980). Comments on Tn3. In *Plasmids and Transposons,* ed. C. Stuttard and K.R. Rozee, pp. 225–8. Academic Press, New York.

Falkow, S. (1975). *Infectious Multiple Drug Resistance.* Pion Ltd. London.

Foster, T.J. and Kleckner, N. (1980). Properties of drug resistance transposons, with particular reference to Tn10. In *Plasmids and Transposons,* ed. C. Stuttard and K.R. Rozee, pp. 207–25. Academic Press, New York.

deGraaff, J., Crosa, J.H., Heffron, F. and Falkow, S. (1978). Replication of the nonconjugative plasmid RSF1010 in *Escherichia coli* K12. *J. Bacteriol.* **134,** 1117–22.

Grindley, N.D.F. and Kelly, W.S. (1976). Effects of different alleles of the *E. coli* K12 *pol* A gene on the replication of nontransferring plasmids. *Molec. Gen. Genet.* **143,** 311–18.

Grindley, N. and Sherratt, D. (1978). Sequence analysis at IS1 insertion sites: models for transposition. *Cold Spring Harbor Symp. Quant. Biol.* **43,** 1257–61.

Jacoby, G.A. (1979). Plasmids of *Pseudomonas aeruginosa.* In *Pseudomonas Aeruginosa Clinical Manifestations of Infection and Current Therapy,* ed. R.G. Doggett, pp. 272–309. Academic Press, New York.

Johnson, D.A. and Willetts, N.S. (1980). F-derived *tra*$^+$ recombinants. Transfer and transposition properties. In *Plasmids and Transposons,* ed. C. Stuttard and K.R. Rozee, pp. 293–301. Academic Press, New York.

Kopecko, D.J. (1979). Specialized genetic recombination systems in bacteria: their involvement in gene expression and evolution. *Progress in Molecular and Subcellular Biology,* vol. 7. Springer-Verlag, Heidelberg.

Mitsuhashi, S. (Ed.) (1977). R-factor. *Drug Resistance Plasmid.* University of Tokyo Press, Tokyo.

Molin, S. and Nordström, K. (1980). Control of plasmid R1 replication: functions involved in replication, copy number control, incompatibility and switch-off of replication. *J. Bacteriol.* **141,** 111–20.

Novick, R.P., Clowes, R.C., Cohen, S.N., Curtis III, R., Datta, N. and Falkow, S. (1976). Uniform nomenclature for bacterial plasmids: a proposal. *Bacteriol. Rev.* **40,** 168–89.

Richmond, M.H., Bennett, P.M., Choi, C.L., Brown, N., Brunton, J., Grinstead, J. and Wallace, L. (1980). The genetic basis of the spread of β-lactamase synthesis among plasmid carrying bacteria. *Phil. Trans. Roy. Soc. Ser. B* **289,** 349–59.

Roussel, A.F. and Chabbert, Y.A. (1978). Taxonomy and epidemiology of gram-negative bacterial plasmids studied by DNA-DNA filter hybridization in formamide. *J. Gen. Microbiol.* **104,** 269–76.

Schlessinger, D. (Ed.) (1978). I. Extrachromosomal elements. In *Microbiology 1978,* pp. 5–286. American Society for Microbiology, Washington.

Schmidt, L., Watson, J. and Willetts, N. (1980). Genetic analysis of conjugation by RP1. In *Plasmids and Transposons,* ed. C. Stuttard and K.R. Rozee, pp. 287–92. Academic Press, New York.

Shapiro, J.A. (1979). Molecular model for the transposition and replication of bacteriophage Mu and other transposable elements. *Proc. Nat. Acad. Sci. USA* **76,** 1933–7.

Thomas, C.M., Meyer, R. and Helinski, D.R. (1980). Regions of broad host range plasmid RK2 which are essential for replication and maintenance. *J. Bacteriol.* **141,** 213–22.

Tomich, P.K., An, F.Y. and Clewell, D.B. (1980). Properties of erythromycin-inducible transposon Tn917 in *Streptococcus faecalis. J. Bacteriol.* **141,** 1366–74.

5

Susceptibility of the whole bacterial cell

Selective inhibition of bacterial growth by antimicrobial agents requires that three important prerequisites be met. A biologically significant target susceptible to the action of low concentrations of the agent must exist, the antibiotic must be able to penetrate the bacterium to the site of the target and the antibiotic must not be inactivated before reaching and interacting with the target. The degree to which these conditions are achieved determines the level of susceptibility of a bacterial cell to an antibiotic.

The cellular locations of targets of antibiotics vary (see Chapter 2). Most are located either in the cytoplasmic membrane or inside the bacterial cell. To reach these targets all antibiotics must penetrate the cell surface layers external to the cytoplasmic membrane. These include the slime or capsule layers and the cell wall. Cell wall structure can be divided, in general terms, into that of gram-positive and gram-negative bacteria. The structure and properties of these cell wall types are quite different and significantly affect the susceptibility of bacteria to various antimicrobial agents.

Cell surface layers and susceptibility

Layers external to the cell wall (capsules and slime)

Most capsular and slime materials of bacteria are polysaccharides although other materials may occur, particularly polypeptides and nucleic acids. These substances are all polyanions and are hydrophilic materials. It is unlikely that such materials would interfere with the passage of hydrophilic antibiotics unless they acted as ion exchange substances. Under these circumstances capsular and slime material could bind positively charged antibiotics such as aminoglycosides and polymyxins. The effect of binding on bacterial susceptibility is unclear. It could conceivably facilitate interaction of the antibiotic with the cell surface and bring the antibiotic in closer proximity to deeper cell layers. Alternatively, by binding the antibiotic, slime and capsular material might retard entry of the antibiotic to deeper cell layers. In the case of hydrophobic antibiotics the hydrophilic and polyanionic nature of the slime and capsular materials could interfere with antibiotic penetration. The hydrophilicity of the slime or capsules would repulse hydrophobic molecules making penetration to deeper lipophilic layers more difficult.

Although it is possible to speculate on the role of extracellular surface layers on antibiotic susceptibility, clear evidence for a significant effect of these substances is generally lacking. It has been reported that strains of *Pseudomonas aeruginosa* which produce 'mucoid' type colonies due to a slime material composed mainly of alginic acid may be slightly more resistant to some antibiotics such as carbenicillin, polymyxins, flucloxacillin and tobramycin. However, not all investigators have agreed that this differential susceptibility of mucoid cells is so, and the exact effect of slime on antibiotic susceptibility even in this situation has not yet been fully answered.

The general role of capsules and slime on antibiotic susceptibility has not been rigorously examined by investigational techniques. This question is clearly worth consideration as most pathogenic bacteria form some sort of extracellular material while growing in animal or human tissues. However, the methods needed to examine the problem are technically difficult because of the potentially different nature of slimes and capsules between host tissues and bacterial culture medium.

Effects of cell wall structure on antibiotic susceptibility

The composition of bacterial cell walls seem most responsible for the characteristic spectrum of activity of many antibacterial agents. The cell wall of gram-negative bacteria provides an apparent barrier to the penetration of various antibiotics to their target sites. A variety of investigations have supported this conclusion. The barrier function of the outer membrane component of the gram-negative bacterial wall accounts for much of the differential susceptibility seen between gram-negative and gram-positive bacteria. The degree of the apparent barrier function, however, varies among different gram-negative bacteria. Strains of *Neisseria* and *Hemophilus* are generally more susceptible to many antibiotics than are strains of *Escherichia coli* which in turn are more susceptible than strains of indole-positive *Proteus* and *Pseudomonas*. Although it is tempting to ascribe this difference in susceptibility to the permeation barrier of the outer membrane of gram-negative bacteria, other factors may contribute to the differences in resistance. For example, almost all strains of *Pseudomonas aeruginosa* produce an inducible, chromosomally specified β-lactamase which apparently accounts for much of the intrinsic resistance of the organism to β-lactam antibiotics.

In addition to posing a barrier to incoming molecules, the outer membrane of gram-negative bacterial cells is responsible for the failure to release antibiotic inactivating enzymes from the cell into the growth medium. Inactivating enzymes in most gram-negative bacteria are probably located on the external surface of the cytoplasmic membrane or within the periplasmic space.

The cell wall of gram-negative bacteria is a complex structure as shown in the diagrammatic representation portrayed in Fig. 5.1. The outermost layer of the cell wall consists of a lipid bilayer in which various proteins are embedded. Extending from the outer surface of the outer membrane is polysaccharide material which represents the O-specific antigen and core material and which is covalently bound to lipid A forming the lipopolysaccharide (LPS) of the cell envelope. Lipid A, as shown in Fig. 5.1, is part of the outer leaflet of the lipid bilayer in association with a variable quantity of phospholipid. The outer membrane containing a lipid bilayer, represents a permeability barrier for hydrophilic molecules.

A series of proteins are present in the outer membrane. Usually there are three or four 'major' proteins and several less frequent 'minor' proteins. The major proteins have been most studied in *E. coli* and *Salmonella typhimurium*. Present evidence indicates that the entry of hydrophilic compounds of molecular weight smaller than about 650 is through protein pores in the outer membrane formed mainly from the major outer membrane proteins termed porins. Porins are associated with the peptidoglycan and usually isolated as trimers. They form diffusions pores when added to phospholipid–LPS mixtures. Their number varies in different bacteria. For example, in *Salmonella typhimurium* there are three (36 000, 35 000, 34 000 molecular weights), in *E. coli* B only one and in *E. coli* K12, two (1a and 1b). The nomenclature of outer membrane proteins varies among authors; that used here is the nomenclature of Schnaitman. An additional porin is found in *E. coli* K12 when it is lysogenized with phage PA-2. These proteins extend from the outer surface of the outer membrane to the peptidoglycan layer. Porin composition of the

Fig. 5.1. A schematic representation illustrating the possible molecular architecture of the *E. coli* cell envelope. Abbreviations used are: PL, phospholipid; OM, outer membrane; PG, peptidoglycan; PS, periplasmic space; and CM, cytoplasmic membrane. Polysaccharide chains in only some of the LPS molecules are shown. The designation ompA is equivalent to tolG. Note that very little phospholipid is distributed in the outer leaflet of the outer membrane. Reproduced, with permission, from the *Annual Review of Biochemistry*, vol. 47. © 1978 by Annual Reviews Inc.

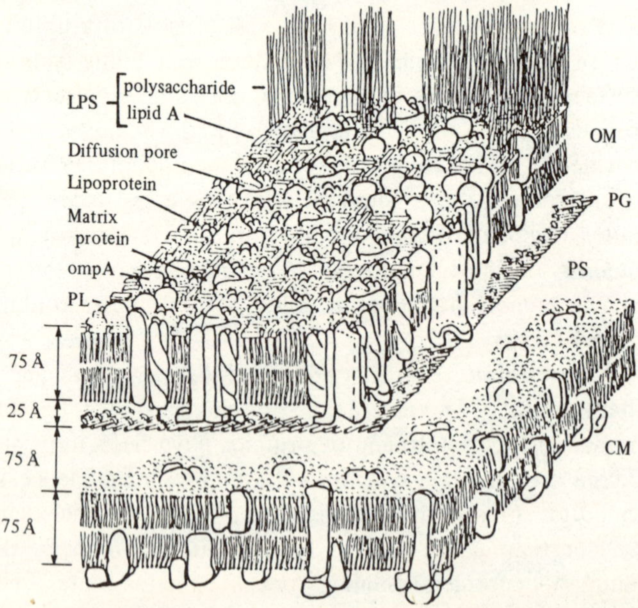

outer membrane is variable and depends, for example, on growth conditions. Porin 1b production is enhanced when cells are grown in trypticase soy broth and 1a production is repressed by salts. Pores are apparently formed from porins in association with the Braun lipoprotein. Lipoprotein is present in both a free form and a form covalently bound to the underlying peptidoglycan layer. This arrangement anchors the porin proteins firmly to the peptidoglycan layer. Pores may be formed from an association of three proteins as shown in Fig. 5.1. Alternatively each porin protein could contain a diffusion pore.

It has been shown that hydrophilic antibiotics like cephaloridine reach the periplasmic space through these pores. A mutation in the structural gene for porin protein 1a (*tol*F) (see Table 5.1) has been shown to cause resistance to several antibiotics including tetracycline, aminoglycosides, chloramphenicol, cephalexin and polymyxins. It has also been shown that in *Neisseria gonorrhoeae* two different resistance loci originally detected in low-level penicillin resistant strains are associated with changes in outer membrane proteins. Locus *mtr*-2 causes resistance to multiple antibiotics and is associated with a seven-fold increase in a minor outer membrane protein as well as an increase in cross linking of peptidoglycan. Locus *pen*B2 producing low-level penicillin and tetracycline resistance is associated with the disappearance of the principal outer membrane protein and the appearance of a new one.

The exclusion limit of 650 molecular weight may not be the case of all bacteria. The limit in *P. aeruginosa* is larger extending to 9000 molecular weight. Although the exclusion limit of 600–650 for Enterobacteriaceae is common, there is evidence that a larger exclusion limit like that of *P. aeruginosa* may be common in other bacteria.

In addition to the relatively non-specific permeability effect of major outer membrane porin proteins on hydrophilic compounds more specific membrane proteins exist. These are involved in the uptake of vitamin B_{12}, iron-chelates, nucleosides and maltose (see Table 5.1) The *ton*A mutation which affects one of the minor membrane proteins causes resistance to the iron containing antibiotic, albomycin. The protein affected by the *ton*A mutation

Table 5.1. *Outer membrane proteins of* Escherichia coli *K12*

Protein	Characteristics	Gene locus
1a	porin protein	*tol*F.*omp*B is a regulatory gene for proteins 1a and 1b.
1b	porin protein	*par*
2	porin protein	Induced by lysogeny with phage PA-2
3a	heat modifiable, required for conjugation	*omp*A
3b	heat modifiable, only present with growth above 37 °C	
Lipoprotein	Braun lipoprotein, present in many Enterobacteriaceae. Present as bound (to peptidoglycan) and free form	*lpp*
Specific diffusion proteins	maltose and maltodextrins, λ receptor	*lam*B
	ferrichrome, T5,T1, colicin receptor	*ton*A
	ferric–enterochelin	*feu*B
	ferric–citrate	*cit*
	ferric–complex	*cir*
	nucleosides	*tsx*
	vitamin B_{12}	*bfe*
Others	products of *tra*S, *tra*T genes	

is involved in iron-ferrichrome uptake. Minor membrane proteins are also, in some instances, receptors for specific bacteriocins which are antibacterial proteins.

Several studies utilizing ‘rough’ mutants of gram-negative bacteria have been used to study the role of lipopolysaccharide on antibiotic susceptibility. Rough mutants are missing a variable part of the polysaccharide portion of the lipopolysaccharide structure of the outer membrane. It has been suggested that polysaccharide material may prevent access of lipophilic antibiotics to the lipid component of the outer membrane bilayer and thus render gram-negative cells more resistant to these antibiotics. Rough mutants are more susceptible to certain antibiotics than their smooth parents. These include macrolides (for example, erythromycin), lincomycins, vancomycin, rifamycin SV, bacitracin, novobiocin, polymyxins, some semisynthetic penicillins (including oxacillin, nafcillin, methicillin, cloxacillin) and actinomycin D.

In studying mutants in which the defect in the polysaccharide extends well into the core region and which are termed 'deep rough' it has been found that resistance to certain antibiotics has increased. For exmple, deep rough lipopolysaccharide mutants are more resistant to benzylpenicillin G, ampicillin, cephaloridine and tetracycline than less rough strains and parental strains. The probable reason for increased resistance is that deep rough mutants undergo an extensive disorganization of the outer membrane and have lost a substantial portion of outer membrane protein. Under these conditions it is probable that the outer membrane protein pores used for entry of drugs like benzylpenicillin G are less available to the drug for permeation to the periplasm.

Studies using several approaches have allowed some tentative conclusions about antibiotics and the outer membrane. These studies include: use of rough mutants of *S. typhimurium,* use of mutants of outer membrane proteins, EDTA treatment of bacteria, susceptibility of spheroplasts versus whole cells, comparison of antibiotic susceptibility of gram-positive and gram-negative bacteria, and comparison of rates of hydrolysis of *ß*-lactam antibiotics in whole and broken cells taking into account the fact that hydrolysis and diffusion rates across the outer membrane are balanced at the steady state.

The outer membrane clearly acts as a barrier to entry of antibiotics. Two general pathways exist to allow diffusion of small molecules across the outer membrane. One is a hydrophilic pathway used by hydrophilic antibiotics (see Table 5.2). This pathway is little affected by the structure of LPS and involves diffusion through water-filled outer membrane pores.

Diffusion rates by the hydrophilic route are influenced by several factors. Rate depends on the size of the molecule and an upper limit exists which, as noted, varies among different bacteria. Increasing hydrophobicity of antibiotics is generally associated

Table 5.2. *Partition coefficients of various antibiotics*

Antibiotic	Partition coefficient
Actinomycin D	>20
Novobiocin	>20
Phenol	>20
Chloramphenicol	12.4
Crystal violet	14.4
Rifamycin SV	8.8
Malachitegreen	4.2
Nalidixic acid	3.16
Minocycline	1.2
Deoxycholate	1.09
Erythromycin	0.79
Clindamycin	0.7
Nafcillin	0.31
Chlortetracycline	0.31
Kanamycin	0.16
Bacitracin	0.12
Oxytetracycline	0.09
Oxacillin	0.07
Oleandomycin	0.07
Tetracycline	0.07
Sodium dodecylsulfate, methylene blue, Penicillin G, cloxacillin	0.02
Neomycin, vancomycin, cycloserine	<0.01
Cepalothin, carbenicillin, ampicillin	<0.01

Partition was determined by equal volumes of octanol and 0.05 M sodium phosphate buffer pH 7.0 at 24 °C. Values of less than 0.02 are clearly hydrophilic. Kanamycin coefficient may be high.

with reduced rates of diffusion among β-lactam antibiotics. A third factor is electrical charge. Anionic cephalosporins penetrate more slowly than the zwitter ion cephalosporin, cephaloridine. This may result from ionic interactions with porins or more likely due to the large number of negatively charged macromolecules in the periplasmic space.

The hydrophilic diffusion route can also change depending on factors which alter the types of protein in the outer membrane. Culture conditions and phage infection can change proteins (see Table 5.1). R-plasmids may reduce the porin content of cells or select for porin-deficient strains and enhance the level of resistance associated with enzymatic modification of an antibiotic. Some bacteria, particularly *H. influenzae,* apparently allow more rapid diffusion of hydrophilic antibiotics perhaps due to larger diameter pores although the mechanism is unknown. The effect is to reduce levels of ampicillin resistance mediated by β-lactamases in such bacteria.

Surprisingly activity of some hydrophobic antibiotics is markedly reduced by mutations of porin 1a. Thus chloramphenicol and the much less hydrophobic agent, tetracycline, appear to use the diffusion pores, at least in part, to cross the outer membrane.

The second major non-specific diffusion pathway is used by many hydrophobic antibiotics. Compounds whose activity is increased in rough mutants and EDTA-treated cells are usually hydrophobic. These agents apparently solubilize particularly in regions of the outer membrane where phospholipids are present in the outer leaflet of the outer membrane. The presence of intact LPS causes this diffusion route to have low activity. The hydrophilic polysaccharide chains of LPS prevent access of hydrophobic molecules to the lipid bilayer. According to these conclusions bacteria with reduced LPS and increased phospolipid in the outer leaflet should be more susceptible to hydrophobic antibiotics. LPS is reduced, for example, in mutations of *E. coli* which are associated with increased susceptibility to hydrophobic antibiotics. Mutations of *acr*A and *rfa*D loci of *E. coli* K12 have decreased LPS and enhanced susceptibility to many hydrophobic agents. The *acr*A mutation tended to increase activity of cationic hydrophobic agents and had a marked decrease in the phosphate content of the lipid A region. Mutants of *rfa*D locus were more susceptible to anionic hydrophobic agents. They had D-glycero-D-mannoheptose in place of L-glycero-D-mannoheptose and had few distal sugars in the LPS molecule. The *acr*A and *rfa*D mutations did change susceptibility slightly for some hydrophilic antibiotics including ampicillin.

Vancomycin is a hydrophilic antibiotic but its activity is enhanced in rough mutants of *S. typhimurium.* The large size of this drug (molecular weight about 3300) may prevent diffusion through the pores.

The relationship between structural differences within a group

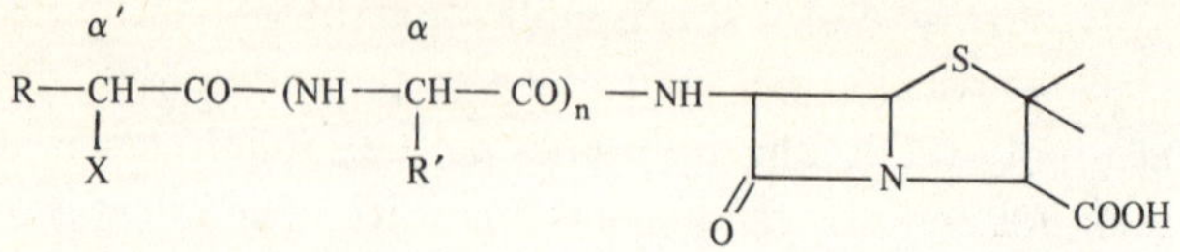

Fig. 5.2. Substituted acetic acid derivatives of 6-amino penicillanic acid. n = 0 or n = 1; R, R′ = various alkyl, aryl, aralkyl and heterocyclic substitutions; X = amino, carboxyl, ureido, SO_3H or combinations of these groups.

of antibiotics and capability to permeate the outer membrane is presently unknown. However, it has been shown that substitutions at the α and α′ positions of the penicillin structure shown in Fig. 5.2 change the spectrum of activity for different gram-positive and gram-negative bacteria. Size, shape and spatial disposition of the substitutions were the important parameters in influencing activity rather than the lipophilic or electronic character of the substitution. Although these substitutions may enhance the interaction with penicillin-sensitive enzymes, it is more likely they influence permeation through the outer membrane. This is probable because the greatest change in activity due to these substitutions occurred for gram-negative bacteria.

Polymyxin antibiotics interact with the outer membrane of many types of gram-negative bacteria. As a result surface blebs and release of periplasmic enzymes occur, due to disruption of parts of the outer membrane. This type of interaction is reduced by divalent cations particularly calcium and magnesium which appear to prevent binding of polymyxins to the polar components of outer membrane phospholipids. This accounts for the antagonistic action of cations on the activity of polymyxins.

An initial interaction with the outer membrane by polymyxin is apparently a necessary prerequisite to the action of these drugs on the cytoplasmic membrane. Bacteria like *Proteus mirabilis* are resistant to polymyxins. If converted to spheroplasts, strains of *P. mirabilis* are 400 times more sensitive to polymyxin B. These and other experiments demonstrate that the cell wall acts as a barrier to the action of polymyxins on the cytoplasmic membrane. It is likely the barrier to polymyxins is part of the outer membrane structure rather than the peptidoglycan which is also removed in spheroplasts. However, not all resistance in bacteria to polymyxins is due to the barrier role of the cell wall. Strains of *Streptococcus fecalis* and *Pseudomonas cepacia* are resistant to polymyxins and cannot be rendered sensitive by converting them to spheroplasts. In bacteria like this it appears that the cytoplasmic membrane is responsible for resistance.

In order for polymyxins to bind to either the outer membrane or the cytoplasmic membrane, negatively charged amphipathic molecules such as cardiolipin are required. There is no evidence that polymyxins interact with only a single type of phospholipid. Most evidence indicates that peptide amino groups bind electrostatically with phospholipid phosphates and the fatty acid tail of the antibiotic penetrates the hydrophobic portion of the membrane. Studies on strains of *P. aeruginosa* resistant to polymyxin have shown a loss in amount of lipopolysaccharide and outer membrane proteins. In these strains resistance is probably the result of a failure of polymyxin to penetrate the outer membrane.

Ethylenediaminetetracetate (EDTA) has been used by many investigators to enhance the activity of several types of antibiotics with a variety of types of bacteria. Many of these experiments have been interpreted to indicate that EDTA reduces the penetration barrier afforded by the outer membrane by causing loss of LPS in gram-negative bacteria. In many instances this conclusion is undoubtedly correct. However, EDTA as a chelating agent is not specific for the outer membrane and certain bacteria are very sensitive to its action (for example *Pseudomonas aeruginosa*). EDTA may also chelate cations from the cytoplasmic membrane and impair that permeability barrier. Thus the specificity of EDTA enhancement of antibiotic activity as an indicator of the role of the outer membrane is often open to doubt. Conclusions based on such experiments, even for penicillin, should be guarded.

The peptidoglycan layer of the cell envelope lies immediately external to the cytoplasmic membrane. It is generally considered to represent less of a penetration barrier than the outer membrane. This conclusion is based mainly on the observation that the peptidoglycan is thick and is the predominant layer in gram-positive bacteria. In spite of this, gram-positive bacteria are generally more susceptible to most antibiotics than gram-negative bacteria. Evidence has been advanced that the activity of certain antibiotics (streptomycin, tetracycline, penicillin) is enhanced in gram-negative bacteria by disruption of the peptidoglycan layer. Whether this is due directly to an effect such as ion exchange provided by the peptidoglycan layer or due to secondary disruption of the whole cell envelope including the outer membrane is not clear at the present time.

Effects of the cytoplasmic membrane on susceptibility of bacteria to antibacterial agents

The targets of some antibacterial agents are in the cytoplasmic membrane. These targets are apparently available to allow binding of the antibiotic without the necessity of a specific mechanism to facilitate entry or transport of the agent into the membrane. Generally speaking penicillin, cephalosporins, vancomycin, ristocetin, bacitracin, polymyxin, cationic detergents, ionophores and uncoupling agents do not require transport mechanisms to cross the cytoplasmic membrane. Other agents are lipophilic at neutral pH and can probably pass through the membrane by solubilization and diffusion. Included in this group are chloramphenicol, trimethoprim, minocycline, fusidic acid, nalidixic acid, rifampicin, novobiocin and probably rosaramicin. For other antibiotics the cytoplasmic membrane is a significant or almost absolute barrier to the target. Included in the last group are the aminoglycosides, aminocyclitols, phosphonomycin, D-cycloserine, tetracycline and perhaps erythromycin, lincomycins and sulfonamides.

D-cycloserine reaches its target using a D-alanine-glycine transport system in *E. coli.* Absence of the components of this transport system results in about an 80-fold increase in resistance to D-cycloserine. Phosphonomycin is transported using transport systems for L-α-glycerophosphate or hexose phosphates. Bacteria are more resistant to phosphonomycin in medium containing glucose and phosphate. These compounds apparently repress or inhibit the L-α-glycerophosphate transport system, accounting for the reduced susceptibility to phosphonomycin. Mutants resistant to phosphonomycin are readily isolated and are most frequently defective in either L-α-glycerophosphate or hexose phosphate transport. D-cycloserine, phosphonomycin, arsenate (using a phosphate transport system) and albomycin (using the *ton*A product involved in ferrichrome uptake) are clear examples of antibiotic agents which cross the cytoplasmic membrane by the fortuitous use of normal transport systems. The mechanism by which other agents cross this barrier is less clear. Two agents, aminoglycosides and tetracyclines, have been studied in considerable depth.

The mechanism mediating aminoglycosides like streptomycin and gentamicin across the cytoplasmic membrane (Fig. 5.3) is

Fig. 5.3. Bacterial entry of aminoglycosides (Ag). Aminoglycosides but not aminocyclitols are shown as $^{+}Ag^{+}$. The polycationic agents diffuse through outer membrane pores in gram-negative bacteria. The process of transfer across the cytoplasmic membrane is similar in both gram-positive and gram-negative bacteria. The transporter is most likely a component of the cytoplasmic membrane intimately associated with creation of the proton gradient. It is either linked to or contains respiratory quinones. Once bound to the transporter, polycationic $^{+}Ag^{+}$ is driven across the membrane by $\Delta\psi$. (EDP-I – see Fig. 5.4). It is bound to ribosomes involved in protein synthesis which act as a 'sink' removing the drug from the membrane. At this time the rate of transport is accelerated (EDP-II – see Fig. 5.4).

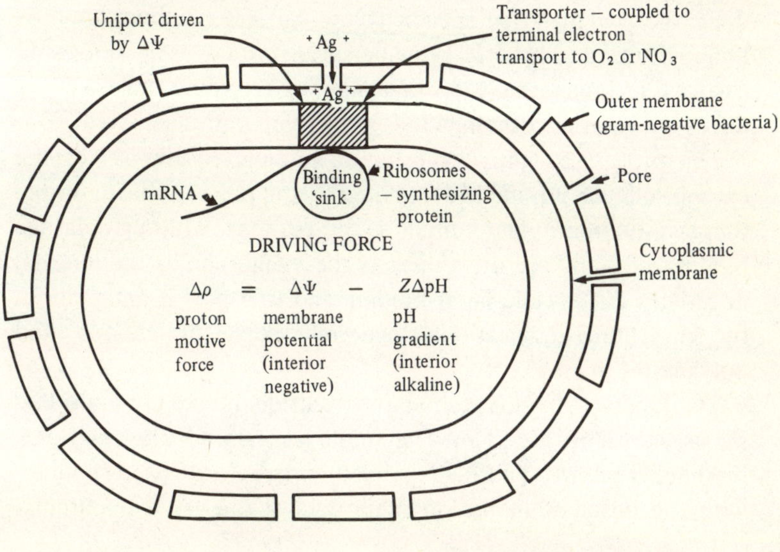

extremely important in determining bacterial susceptibility to these agents. The initial entry of streptomycin and gentamicin into bacteria is energy dependent with transport being dependent on a protonmotive force.

The cellular protonmotive force ($\Delta\rho$) is composed of two components. These are a cross-membrane electrical potential ($\Delta\psi$) and pH gradient (ΔpH) with the interior of the cell being negative and alkaline at neutral pH. The positively charged aminoglycosides are driven across the membrane mainly by the magnitude of $\Delta\psi$.

The most effective energy source for the bacterial cell to cause aminoglycoside uptake is electron transport to oxygen, with electron transport to nitrate being somewhat less effective. Electron transport to fumarate or ATP from substrate phosphorylations are very inefficient energy sources. The differential importance of energy sources is probably because aminoglycosides bind to a membrane component (transporter) involved in membrane energization from electron transport. There is evidence that respiratory quinones are needed for binding either directly as binding sites or indirectly in that they are needed to create a binding site or to create the necessary $\Delta\psi$ to drive aminoglycoside entry. Binding sites could be a series of carrier proteins binding aminoglycosides on the basis of their polycationic nature. However, the evidence that specific solute carriers are 'borrowed' by aminoglycosides is poor. The evidence against the use of borrowed carriers includes the fact that aminoglycoside uptake shows diffusion kinetics although it is clearly energy dependent.

The process of initial transfer of aminoglycosides across the cytoplasmic membrane is dominated by the polycationic nature of these antibiotics. They bind as polycations with cytoplasmic binding sites and are driven across the membrane by the internal negativity of the cell. Their binding and transport is antagonized by competing polycationic especially divalent cations and polyamines.

The kinetics of whole cell aminoglycoside uptake are multiphasic as shown in Fig. 5.4. The initial very rapid phase is ionic binding. The first energy-dependent phase (EDP-I) terminates with the onset of inhibition of protein synthesis. It apparently represents the initial energy-requiring transfer of the aminoglycoside across the cytoplasmic membrane to the ribosome.

The second energy-dependent phase (EDP-II) has been shown to require ribosomal binding of streptomycin and gentamicin. It is characterized by a significant increase in the rate of entry. However, it has identical energy requirements to EDP-I.

The basis of EDP-II is not fully understood. Unquestionably interaction of drug with ribosomes involved in protein synthesis is required. The induction of a generalized polyamine transport to transport aminoglycosides has been proposed. Although streptomycin competes with spermidine and putrescine uptake, polyamines do not induce EDP-II. Uptake of polyamines and aminoglycosides can be dissociated in some mutants and streptomycin and gentamicin do not show saturable kinetics during EDP-II uptake. These and other points are strong evidence against this proposal.

EDP-II is more likely due to some combination of two sequelae of ribosomal binding of aminoglycosides. At the time of onset of EDP-II, there is a marked loss of K^+ from cells which would increase the magnitude of $\Delta\psi$. This could explain the increased rate of uptake of arginine, polyamines and other aminoglycosides that occurs during EDP-II. In addition the changes in ribosomal cycling seen with streptomycin binding may increase the effectiveness of a ribosomal 'sink' effect and increase the rate of binding aminoglycosides and thus removal from the membrane.

Fig. 5.4. Kinetics of uptake of gentamicin and streptomycin.

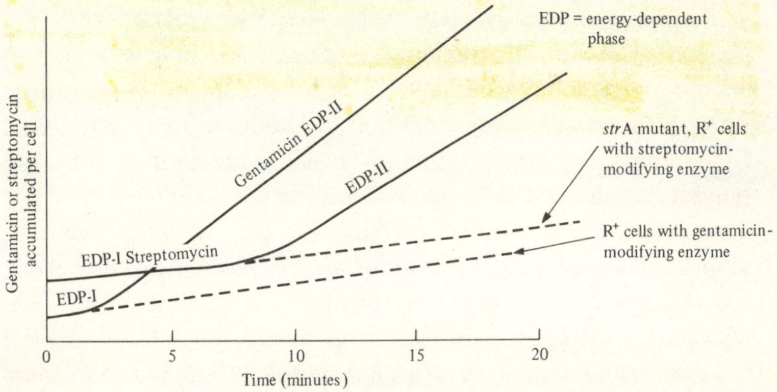

This is supported by the necessity for ribosomal binding to initiate EDP-II and by the observation that puromycin enhances streptomycin uptake.

The nature of the energy sources which are needed for EDP-I determines much of the susceptibility of bacteria to aminoglycosides. Bacteria which can carry out oxygen- or nitrate-dependent respiration coupled to oxidative phosphorylation are, in the absence of additional resistance mechanisms, susceptible to streptomycin, gentamicin and probably other aminoglycosides in general. Other bacteria like members of the genus *Clostridium* which do not carry out electron transport to the preceding carriers or like *Bacteroides fragilis* which only reduce fumarate have a very low susceptibility to aminoglycosides. Primarily fermentative bacteria, like all streptococci, are relatively resistant. Anaerobic bacteria that reduce nitrate, like *Bacteroides ureolyticus,* are sensitive to aminoglycosides. Facultative bacteria such as *E. coli* or *Staphylococcus aureus* are much more resistant to aminoglycosides when grown fermentatively than when reducing oxygen by electron transport. *Pseudomonas aeruginosa* is a strict aerobe and can only grow anaerobically by reduction of nitrate or nitrite. Under the latter condition strains are about 4–8 times as resistant to gentamicin or tobramycin then when grown with oxygen.

Divalent cations and, to a considerably less extent, monovalent cations antagonize the action of aminoglycosides. This effect is so in most, if not all, bacteria but is especially pronounced in *Pseudomonas aeruginosa.* Cation antagonism is most likely exerted in two ways. Cations inhibit uptake of aminoglycosides in spheroplasts as well as whole cells. Therefore it is probable that cations compete with aminoglycosides for binding sites represented by polar phospholipid heads and perhaps respiratory quinones in the cytoplasmic membrane. In addition in *P. aeruginosa* the outer membrane contains greater substitution of core polysaccharides by phosphate residues than is so in most bacteria. This seems an additional binding site for which cations and aminoglycosides compete. Both of these effects of cations reduce the amount of aminoglycoside available to undergo membrane transport and increase resistance to aminoglycosides.

Tetracycline uptake has been extensively studied and some

portion of the uptake is energy dependent at least in the case of tetracycline and, likely, oxytetracycline. Levy and his colleagues have reported that tetracycline uptake in *E. coli* involves both energy-independent and energy-dependent components. Uptake of tetracycline occurs by an initial rapid energy-independent mechanism. This process reaches equilibrium within a few minutes and a slower continuing energy-dependent phase can then be detected. The latter uptake is sensitive to inhibitors of electron transport and energy coupling and to uncoupling mutations. Thus tetracycline may enter during the energy-dependent phase by using energy from either electron transport or ATP formed from electron transport. Glycolytic ATP does not seem to be effective as an energy source. Levy and colleagues have noted that this unusual energy requirement needs further confirmation. The initial rapid energy-independent uptake explains why anaerobic bacteria are susceptible to tetracycline. Under anaerobic conditions when the slow uptake phase is inhibited, tetracycline still accumulates by the energy-independent mechanism and inhibits cell growth. The two phase uptake also explains the difficulty in isolating transport mutants, as a single-step mutation would abolish only one transport system.

The transport of tetracycline has differences from the transport of aminoglycosides like streptomycin and gentamicin which account for differences in the antibacterial spectrum between these two major groups of antibiotics. Aminoglycosides tested to date most effectively utilize a transporter and driving force developed from terminal electron transport for initial bacterial entry. Intrinsic susceptibility in bacteria is determined in large part by the extent of this process. Thus, for example, anaerobic bacteria which do not reduce nitrate are highly resistant. In contrast, anaerobic bacteria of this type are often very sensitive to tetracycline. Presumably this difference is due to the energy-independent transport of tetracycline. Similarly streptococci are relatively resistant to aminoglycosides but generally sensitive to tetracycline (in the absence of an acquired resistance mechanism). Facultative bacteria growing fermentatively are fully susceptible to tetracycline but are relatively resistant to aminoglycosides. Characteristics of the transport processes of these two groups of drugs thus account for much of the characteristic susceptibility of different groups of bacteria to the two groups of agents.

In the case of the tetracycline derivative minocycline, it is probable that the cytoplasmic membrane is crossed by solubilization of the drug and subsequent diffusion across the lipid bilayer. Minocycline is very much more lipophilic than tetracycline. Doxyclycline is intermediate in its lipophilicity and the manner by which it crosses the cytoplasmic membrane is less clear.

Tetracycline uptake has been shown to occur by light-induced carrier mediated transport in *Rhodopseudomonas sphaeroides*. Efflux and influx of the drug were mediated by the system. An energy-independent system was not described.

The inter-relationship between resistant mechanisms and other factors influencing antibiotic susceptibility of the whole cell

The function of resistance mechanisms in the whole and living bacterial cell has been studied particularly for penicillins and aminoglycosides. Other antibiotics have been studied to a lesser degree and the whole cell relationship is not well enough understood to construct a clear picture.

Resistance to β-lactam antibiotics resulting from β-lactamase enzyme activity depends on the nature of the bacterial cell wall. In gram-positive bacteria the barrier to inflow of the antibiotic or outflow of the β-lactamase enzyme is relatively small (see Fig. 5.5 part *a*). The β-lactamase produced by members of the population is free to enter the environment and in most instances is diluted in the external environment. Resistance is achieved by the progressive destruction of the β-lactam antibiotic in the extracellular milieu. When the concentration of the antibiotic falls below a critical level, inhibition of penicillin- or cephalosporin-sensitive enzymes involved in cell wall synthesis is inadequate to prevent cell growth. The external growth medium is detoxified (Fig. 5.5 *a*ii). Under these circumstances resistance is a population phenomenon. The total β-lactamase activity produced by all members of the bacterial population acts in a cumulative fashion to destroy the β-lactam antibiotic. Therefore, the larger the inoculum of bacteria, the higher is resistance because the antibiotic is degraded

Fig. 5.5. Whole cell β-lactamase mediated resistance to β-lactam antibiotics.

(a) Gram-positive bacteria

(i) *No additional β-lactamase*

β-lactam in external medium → Diffuses through cell wall → Binds to PBP → Inhibition of peptidoglycan; → Activation of autolytic enzymes

(ii) *With β-lactamase*

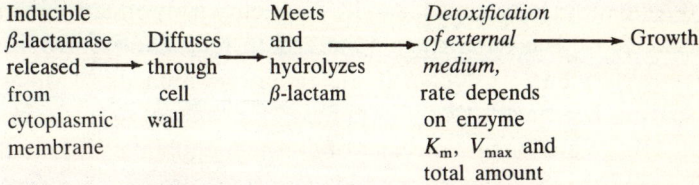

Inducible β-lactamase released from cytoplasmic membrane → Diffuses through cell wall → Meets and hydrolyzes β-lactam → *Detoxification of external medium,* rate depends on enzyme K_m, V_{max} and total amount → Growth

(b) Gram-negative bacteria

(i) *No additional β-lactamases*

β-lactam in external medium → Diffuses through outer membrane → Periplasm → Binds to PBPs → Various morphological effects, growth inhibition, ± death

(ii) *Additional β-lactamase – 'poorly permeable β-lactam'*

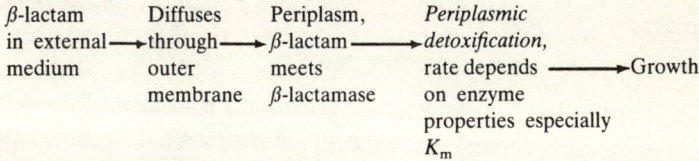

β-lactam in external medium → Diffuses through outer membrane → Periplasm, β-lactam meets β-lactamase → *Periplasmic detoxification,* rate depends on enzyme properties especially K_m → Growth

(iii) *Additional β-lactamse, permeable β-lactam or outer membrane*

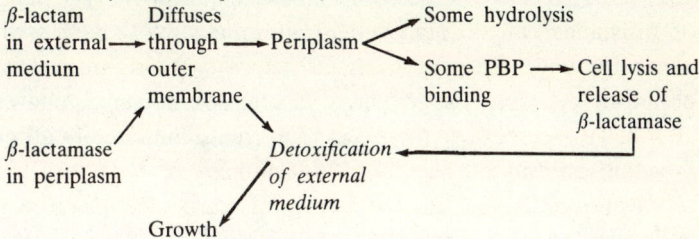

β-lactam in external medium → Diffuses through outer membrane → Periplasm → Some hydrolysis; → Some PBP binding → Cell lysis and release of β-lactamase → *Detoxification of external medium*

β-lactamase in periplasm → Diffuses through outer membrane → *Detoxification of external medium* → Growth

---

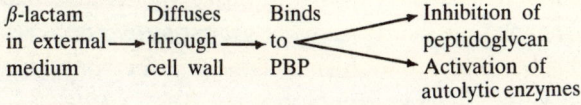

more rapidly and completely. Gram-positive bacteria are usually very susceptible to the action of β-lactam antibiotics because they posses little in the way of a cell wall barrier and the targets are highly susceptible to binding the antibiotic. This combination of release of β-lactamases and the high susceptibility to β-lactam antibiotics means that β-lactamases of gram-positive bacteria are usually inducible and produced in relatively large amounts with high affinities for β-lactam antibiotics.

In many gram-negative bacteria β-lactam resistance from β-lactamases is achieved in a different but highly effective manner (Fig. 5.5). The outer membrane acts to retard entry of β-lactam antibiotics and β-lactamases are retained within the periplasmic space. The effect is that β-lactamases are situated so that they can inactivate incoming molecules of β-lactam antibiotic prior to their binding to the β-lactam antibiotic-sensitive enzymes (transpeptidases, D, D-carboxypeptidases, endopeptidases) of cell wall synthesis. In contrast to the situation described above, only periplasmic β-lactam needs detoxification rather than that of the whole medium. Each cell acts as a relatively self-contained resistance unit (Fig. 5.5 part *b*ii).

Resistance is an interplay of the properties of the enzyme present in a particular bacterium (K_m, V_{max}, total enzyme per cell) which determines the rate of hydrolysis of the β-lactam at a specific concentration of the β-lactam and the rate of entry of that drug to the periplasm of gram-negative bacteria. From this it follows that the level of resistance to two penicillins or cephalosporins cannot be accurately forecast by knowing only the properties of the enzyme. This is an efficient resistance mechanism. Very high levels of resistance can be achieved by gram-negative bacteria which generally have lower β-lactamase activity than gram-positive organisms. R-factors specifying β-lactamases may also apparently alter cell envelope structure so as to decrease the rate of entry of a β-lactam antibiotic. The TEM β-lactamase of R-factor RP1 is relatively inefficient at hydrolysis of carbenicillin (10% the hydrolysis rate of penicillin) but produces high resistance to carbenicillin. Current evidence supports the view that RP1 also modifies the cell wall and probably decreases the rate of entry of carbenicillin. This results in the equilibrium shifting towards resistance as more external carbenicillin is needed to produce a rate of entry which exceeds the rate of destruction of the drug (Fig. 5.5 part *b*ii).

Cephaloridine is a zwitterion cephalosporin and diffuses through the outer membrane of gram-negative bacteria more rapidly than other most other cephalosporins. It also has a relatively high affinity for penicillin binding protein 1b. The concentration in the periplasm can exceed the capability of a β-lactamase to hydrolyze it. Under these circumstances, cell lysis results, β-lactamase is released to the medium and medium detoxification is required before growth is reinitiated. This situation is similar to that of gram-positive bacteria and in 'highly-permeable' bacteria like *H. influenzae*. (Fig. 5.5 part *b*iii).

Some gram-negative bacteria produce relatively low resistance levels for β-lactam antibiotics. Even with a β-lactamase present strains of *Hemophilus influenzae* may have relatively low-level ampicillin resistance in spite of a β-lactamase capable of hydrolyzing ampicillin. This seems to be due to the relatively greater permeability of ampicillin through the outer membrane of the cell wall of *H. influenzae* compared to *E. coli* perhaps due to larger exclusion limits of the pores of *H. influenzae*. Under these circumstances with a high enough external concentration of ampicillin, the capability to hydrolyze ampicillin is exceeded by the rate of entry of the drug and penicillin sensitive enzymes are inhibited by unhydrolyzed ampicillin (Fig. 5.5 part *b*iii).

Resistance to aminoglycosides mediated by aminoglycoside-inactivating enzymes is also dependent on the nature of aminoglycoside entry into bacteria. Only a very few bacteria inactivate a significant fraction of the aminoglycoside (e.g. streptomycin, gentamicin, kanamycin, amikacin) contained in the culture medium. This is true even though such bacteria possess enzymes capable of inactivating specific aminoglycosides using extracts from broken or osmotically-shocked cells. A few strains of *E. coli* possessing streptomycin phosphotransferases have been reported capable of inactivating a significant fraction of extracellular streptomycin. However, the vast majority of gram-negative bacteria inactivate less than 0.5% of the extracellular drug present. In spite of this such bacteria are most often able to grow in the presence of high concentrations of those aminoglycosides for which they have inactivating enzymes.

Evidence has steadily accrued from studies using *E. coli* and *P. aeruginosa* that resistance is the outcome of competition between two rates. These are: (a) The rate of inactivation of the aminoglycoside which in turn depends on the properties of the inactivating enzyme for the agent in question (K_m, V_{max}, amount/cell). The K_m value is of particular importance as it is a measure of affinity of enzyme for substrate. The lower the K_m value, the lower the concentration of aminoglycoside needed to initiate inactivation of the drug. (b) The rate of initial aminoglycoside transport into the cell. The characteristics of transport were noted earlier.

The most comprehensive studies on the interplay of transport and inactivation have been done for streptomycin adenylyltransferase and *E. coli.* However, several studies have shown that the level of whole cell resistance to different aminoglycosides is highly dependent on the relative K_m value of an enzyme for each of the agents. These studies indicate that this is a general mechanism which probably holds for all aminoglycosides (spectinomycin has differing transport characteristics and is not an aminoglycoside).

Streptomycin passes through outer membrane protein pores, reaches the periplasm and binds to polar heads of phospholipids within the cytoplasmic membrane. At this point the drug comes into contact with the inactivating enzyme located on the external surface of the cytoplasmic membrane. Membrane transport sites for streptomycin are likely located in near proximity to the inactivating enzymes. The nature of these has been noted earlier in the chapter (see Fig. 5.3). In the presence of concentrations of streptomycin where the rate of initial streptomycin entry does not exceed the rate of streptomycin adenylylation, all streptomycin reaching the cytoplasmic membrane is adenylylated and transported as the adenylylate. Acetate and phosphate derivatives produced by aminoglycoside acetylating or phosphorylating enzymes are likely similarly transported. These inactivated molecules do not bind to ribosomes and the sequelae of the drug action do not develop. The enhanced rate of uptake of streptomycin, which normally occurs after a variable period of streptomycin treatment in streptomycin-sensitive *E. coli,* does not occur demonstrating the dependence on ribosomal binding of this phase of uptake (Fig. 5.4). After 30–60 minutes of transport of the inactivated streptomycin, the initial accumulation rate declines and eventually ceases, probably because available transport and cellular binding sites are occupied at that particular concentration of streptomycin. This occurs when about $\frac{1}{2}$% of the total drug is adenylylated. Thus resistance is achieved by inactivating only that fraction of drug undergoing transport. This mechanism results in bacterial cells growing in a milieu containing active drug.

The preceding concept of resistance can be termed the 'rate competition' model. It is supported by the fact that mutants with reduced streptomycin and other aminoglycoside transport rates and which contain R-factor mediated inactivating enzymes (R^+) have higher resistance levels than parent R^+ bacteria. This observation is explained by the fact that a higher external drug concentration is needed for the mutant to produce a transport rate that equals or exceeds the rate of inactivation. The reverse effect is found with mutants that have higher transport rates. The rate competition model is also supported by observations obtained from study of the inactivating enzymes. Aminoglycoside-inactivating enzymes result in highest resistance levels to those aminoglycosides for which they have the lowest K_m and highest V_{max} values. For example, several inactivating enzymes produce resistance to gentamicin or sisomicin but are more susceptible to the 5-epi-derivatives of the two drugs. This is associated with significantly lower K_m values for gentamicin and sisomicin than for the derivatives. In the case of the 5-epi-derivatives, transport is likely underway at concentrations below those needed to initiate significant inactivation.

When the external aminoglycoside concentration is raised, the rate of initial transport is increased and eventually exceeds the rate of inactivation. Under these conditions active drug is transported, binds to ribosomes and misreading of mRNA and/or inhibition of protein synthesis occur as well as modification of membrane permeability. This concept of the mechanism of aminoglycoside resistance in bacteria possessing aminoglycoside-inactivating enzymes explains the occasional observation that some aminoglycoside-inactivating enzyme will inactivate an aminoglycoside *in vitro* but resistance is not obtained in the whole cell. This is apparently true because the binding affinity (K_m) for the aminoglycoside in question is higher than effective concentrations obtained in the periplasm. Alternatively the aminoglycoside may be rapidly transported. The characteristics of the aminoglycoside-inactivating

Table 5.3. *Resistance mechanisms operative in whole cells*

Mechanism		Examples of antibiotics exhibiting the mechanism
(A) Alteration of target		
(i) various mutations		streptomycin, spectinomycin, fusidic acid, kasugamycin, rifampicin, nalidixic acid, novobiocin, low-level penicillin
(ii) plasmid		macrolide-lincomycin-methylation of two 23S RNA adenine residues
(B) Altered drug transport		
(i) various mutations		cycloserine, phosphonomycin, albomycin, arsenate
(ii) plasmid		tetracycline (possibly in combination with modified ribosome?), cadmium,? chloramphenicol and others
(C) Detoxification		
plasmid	(a)	of drug in medium – chloramphenicol (some exceptions) – mercury – β-lactams gram-positive bacteria
	(b)	of drug in periplasm – β-lactams – many gram-negative bacteria
	(c)	periplasm and medium – some β-lactams e.g. cephaloridine – some bacteria e.g. *Hemophilus influenzae*
(D) Bypass enzymes		
plasmid		trimethoprim – new dihydrofolate reductase sulfonamides – new dihydropteroate synthetase
(E) Inactivation of drug undergoing transport		
plasmid		aminoglycosides

enzyme result in a rate of inactivation which is not rapid enough to exceed the rate of transport.

Table 5.3 summarizes most of the resistance mechanisms as they operate in whole cells. The variation by which acquired inactivating enzymes achieve resistance is seen among resistance to β-lactams, chloramphenicol and aminoglycosides.

Selected references

Brown, M.R.W. (1975). The role of the cell envelope in resistance. In *Resistance of Pseudomonas aeruginosa,* M.R.W. Brown, pp. 71–108. John Wiley and Sons, London.

Bryan, L.E. (1980). Mechanisms of plasmid mediated drug resistance. In *Plasmids and Transposons,* ed. C. Stuttard and K.R. Rozee, pp. 57–81. Academic Press, New York.

Bryan, L.E., Kowand, S.M. and Van Den Elzen, H.M. (1979). Mechanisms of aminoglycoside antibiotic resistance in anaerobic bacteria: *Clostridium perfringens* and *Bacteroides fragilis. Antimicrob. Agents Chemother.* **15,** 7–13.

Bryan, L.E. and Van Den Elzen, H.M. (1977). Effects of membrane-energy mutations and cations on streptomycin and gentamicin accumulation by bacteria: a model for entry of streptomycin and gentamicin in susceptible and resistant bacteria. *Antimicrob. Agents Chemother.* **12,** 163–77.

Chopra, I. and Howe, T.G.B. (1978). Bacterial resistance to the tetracyclines. *Microbiol. Reviews* **42,** 707–24.

Coleman, W.G. and Lieve, L. (1979). Two mutations which affect the barrier function of the *Escherichia coli* K-12 outer membrane. *J. Bacteriol.* **139,** 899–910.

Costerton, J.W. and Cheng, J.J. (1975). The role of the bacterial cell envelope in antibiotic resistance. *J. Antimicrob. Chemother.* **1,** 363–77.

Dickie, P., Bryan, L.E. and Pickard, M.A. (1978). Effect of enzymatic adenylylation on dihydrostreptomycin accumulation in *Escherichia coli* carrying an R-factor: model explaining aminoglycoside resistance by inactivating mechanisms. *Antimicrob. Agents Chemother.* **14,** 569–80.

DiRienzo, J.M., Nakamura, K. and Inouye, M. (1978). The outer membrane proteins of gram-negative bacteria: biosynthesis, assembly and functions. *Ann. Rev. Biochem.* **47,** 481–532.

Fayolle, F., Privitera, G. and Sebald, M. (1980). Tetracycline transport in *Bacteroides fragilis. Antimicrob. Agents Chemother.* **18,** 502–5.

Gilleland, H.E. Jr and Conrad, R.S. (1980). Effects of carbon sources on chemical composition of cell envelopes of *pseudomonas aeruginosa* in association with polymyxin resistance. *Antimicrob. Agents Chemother.* **17,** 623–8.

Gilleland, H.E. Jr and Lyle, R.D. (1979). Chemical alterations in cell envelopes of polymyxin-resistant *Pseudomonas aeruginosa* isolates. *J. Bacteriol.* **138,** 839–45.

Govan, J.R.W. and Fyfe, J.A.M. (1978). Mucoid *Pseudomonas aeruginosa* and cystic fibrosis: resistance of the mucoid form to carbenicillin, flucloxacillin and tobramycin and the isolation of mucoid variants *in vitro. J. Antimicrob. Chemother.* **4,** 233–40.

Guyman, L.F., Walstad, D.L. and Sparling. P.F. (1978). Cell envelope alterations in antibiotic sensitive and resistant strains of *Neisseria gonorrhoeae*. *J. Bacteriol.* **136,** 391–401.

Hancock, R.E.W. and Nikaido, H. (1978). Outer membranes of gram-negative bacteria, XIX, Isolation from *Pseudomonas aeruginosa* PAO1 and use in reconstitution and definition of the permeability barrier. *J. Bacteriol.* **136,** 381–90.

Holtje, J.V. (1978). Streptomycin uptake via an inducible polyamine transport system in *Escherichia coli*. *Eur. J. Biochem.* **86,** 345–351.

Kadner, R.J. (1978). Transport of vitamins and antibiotics. In *Bacterial Transport*, ed. B.P. Rosen pp. 463–94. Marcel Dekker Inc., New York.

Levy, S.B. and McMurry, L. (1978). Plasmid-determined tetracycline resistance involves new transport systems for tetracycline. *Nature* (Lond.) **276,** 90–2.

McMurry, L. and Levy, S.B. (1978). Two transport systems for tetracycline in sensitive *Escherichia coli:* critical role for an initial rapid uptake system insensitive to energy inhibitors. *Antimicrob. Agents Chemother.* **14,** 201–9.

Nikaido, H. and Nakae, T. (1979). The outer membrane of gram-negative bacteria. *Advances Microbiol. Physiol.* **20,** 163–250.

Richmond, M.H. (1978). Factors influencing the anti-bacterial action of β-lactam antibiotics. *J. Antimicrob. Chemother.* **4,** 1 – 14.

Rosenberg, H., Gerdes, R.G. and Chegwidden, K. (1977). Two systems for the uptake of phosphate in *Escherichia coli*. *J. Bacteriol.* **131,** 505–11.

Sanderson, K.E., MacAlister, T., Costerton, J.W. and Cheng, K.J. (1974). Permeability of lipopolysaccharide-deficient (rough) mutants of *Salmonella typhimurium* to antıbiotics, lysozyme and other agents. *Canadian J. Microbiol.* **20,** 1135–45.

Schnaitman, C.A. (1974). Outer membrane proteins of *Escherichia coli*. III. Evidence that the major protein of *Escherichia coli* 0111 outer membrane consists of four distinct polypeptide species. *J. Bacteriol.* **118,** 442–53.

Silhavy, T.J., Hartig-Beecken, I. and Boos, W. (1976). Periplasmic protein related to the sn-glycerol-3-phosphate transport system of *Escherichia coli*. *J. Bacteriol.* **126,** 951–8.

Wargel, R.J., Shadur, C.A. and Neuhaus, F.C. (1971). Mechanism of D-cycloserine action: transport mutants for D-alanine, D-cycloserine and glycine. *J. Bacteriol.* **105,** 1028–35.

Weckesser, J. and Magnuson, J.A. (1979). Light-induced, carrier-mediated transport of tetracycline by *Rhodopseudomonas sphaeroides*. *J. Bacteriol.* **138,** 678–83.

Zimmermann, W. and Rosselet, A. (1977). Function of the outer membrane of *Escherichia coli* as a permeability barrier to beta-lactam antibiotics. *Antimicrob. Agents Chemother.* **12,** 368–72.

6

Current problems of resistant bacteria

An understanding of the mechanisms and epidemiology of antibiotic resistance is more than scientific curiosity. Many bacteria have posed and continue to pose therapeutic problems due to antibiotic resistance. The seriousness of these problems varies among bacteria, antibiotics and geographic locations at different times. This chapter considers some of the more important problems of antibiotic resistance for different bacteria.

Staphylococcus aureus and S. epidermidis

Resistance to benzyl penicillin G and related antibiotics is commonly detected in clinically significant isolates of *S. aureus*. In 1979, 84.5% of strains were resistant to penicillin at the Foothills Hospital in Calgary, Canada. Resistance percentages in this range are common in many parts of the world for patients either from the general community or in hospitals.

Penicillin resistance is specified by an inducible penicillinase. Benzylpenicillin and other substrates as well as nonsubstrates like methicillin, cloxacillin and other penicillinase-resistant penicillins are inducers. The compound 2-(2′-carboxyphenyl) benzoyl-6-aminopenicillanic acid (CBAP) is a gratuitous inducer acting as neither substrate nor effective antibiotic. Inducers may activate constitutively synthesized antirepressor protein (see Chapter 3). In turn this compound may complex with the repressor (pen-I gene) and allow synthesis of penicillinase (pen-P gene). The control of all penicillinases may not be identical (see Chapter 3).

Clinically important penicillin resistance is almost always specified by plasmids reported to range in size from 12×10^6 to 21×10^6. Four serotypes of penicillinase have been described. Other resistances have been linked with penicillin on the same plasmid. These include: resistance to cadmium in particular but also erythromycin, kanamycin, fusidic acid and other heavy metals. Plasmid transfer normally occurs by transduction although transformation of plasmids has been described and may occur in nature.

Resistance to methicillin and other penicillins resistant to staphylococcal *β*-lactamase ('methicillin resistance') is uncommon in most locations (0.1% at the Foothills Hospital). Higher frequencies have been reported occasionally particularly in Europe (45% at one time in Denmark, now much lower) and somewhat more frequently recently in the US. Testing at reduced temperature or with NaCl in the medium may be needed to detect this form of antibiotic resistance as only a small minority of the population may be resistant at 37 °C with unsupplemented medium.

Strains resistant to oxacillin in the absence of added NaCl at 37 °C have been reported including isolates from serious and fatal infections. In these strains it has been suggested that a higher proportion of resistant cells exists than in most methicillin-resistant isolates. However, relatively few isolates from serious staphylococcal infections are resistant to methicillin and related drugs. Methicillin-resistant strains have been reported to be eliminated from burns by flucloxacillin. The therapeutic response of methicillin-resistant *S. aureus* to penicillinase-resistant penicillins or cephalothins remains unclear. Such strains currently pose a relatively small therapeutic problem because of their low prevalence.

The majority of methicillin-resistant strains also show reduced susceptibility to cephalothin and some other cephalosporins. This is particularly marked if minimal bactericidal concentrations (MBCs) are measured. It is probably wise to consider that all methicillin-resistant strains have reduced susceptibility to most cephalosporins particularly for therapy of endocarditis. Treatment of infections due to methicillin-resistant *S. aureus* with cephalothin has been associated with poor clinical results on several occasions.

S. epidermidis may be the cause of infections (endocarditis, cerebrospinal fluid, shunt infections, hip infections) particularly associated with a prosthetic device. The percentage of methicillin-resistant strains is usually higher than for *S. aureus* (e.g. 12% at

Foothills Hospital). Again many of these are relatively resistant to cephalosporins including cephalothin. In spite of resistance to penicillinase-resistant penicillins or cephalosporins, many physicians will use one of these drugs (usually the most active by MBC determinations) combined with an aminoglycoside for therapy of methicillin-resistant *S. epidermidis* infections. Alternative agents include vancomycin and perhaps vancomycin–cephalosporin or vancomycin–rifampicin combinations.

Some strains of *S. aureus* have been reported as tolerant to penicillins. Variation in definition of tolerance by different investigators has occurred. Best and colleagues define tolerance in those strains with a ratio of MBC:MIC of >10 after 24 hours incubation (see Chapter 3 for a discussion of the mechanism). Sabath and colleagues use a MBC:MIC ratio of at least 32 at 24 hours. Tolerance depends on the time of incubation in that some strains with a high MBC:MIC ratio after 8 hours of penicillin treatment may have a ratio of 1 at 18 hours. Thus different rates of killing may ultimately be a better way to define tolerance.

Some patients with endocarditis who have responded poorly to penicillin therapy have had tolerant staphylococci. However, in a study of *S. aureus* endocarditis in rabbits, no difference in response to methicillin could be shown for tolerant and non-tolerant strains. At the present time it is probably important to look for tolerance defined by standardized criteria in patients with endocarditis who have low serum bactericidal activities (trough values of <1:8) or in those who fail to respond rapidly to therapy with penicillins.

Resistance to other antimicrobial agents varies considerably depending on factors such as geographic location, hospital v. non-hospital isolate, underlying disease, hospital unit, types of antibiotics in prevalent use and others. Resistance to erythromycin, lincomycin, clindamycin, tetracycline, fusidic acid and novobiocin has been recognized as clinically significant resistance at various times particularly when agents are used singly. Chloramphenicol and tetracycline resistances are borne on small molecular weight plasmids (about3 $\times 10^6$) present as multiple copies per cell.

Recently gentamicin-resistant strains have been increasingly detected and many of these are multiply-resistant strains. Gentamicin and multiply-resistant strains have been associated with bacteremia and fatal infections. A variety of aminoglycoside-modifying enzymes have been detected in *S. aureus* strains including more than one type in a single strain. At least some of these strains possess plasmids. In one outbreak of multiply-resistant *S. aureus,* infected patients had long hospital stays, developed their infection after several hospital days and had received prior antibiotics.

Treatment of serious staphylococcal sepsis including pneumonia still has a relatively high failure rate. Recently the use of combined antibiotics (a penicillin and gentamicin, a penicillin and fusidic acid, a penicillin and erythromycin) has been initiated. Rifampicin has also been successfully used in some serious *S. aureus* infections. Resistance to rifampicin is readily obtained *in vitro* and the drug probably should not be used as a single agent.

Streptococcus spp.

Streptococcus pneumoniae

Antibiotic-resistant *S. pneumoniae* have been reported for over 15 years but have not until recently posed a serious therapeutic problem. Resistance to tetracyclines was the most frequently reported. However, resistance to erythromycin (high and low level), lincomycin, clindamycin, sulfonamides and chloramphenicol have been described. Penicillin resistance has been recognized for close to 20 years, mainly in New Guinea and Australia.

Recently in South Africa, a large outbreak of strains showing various resistance patterns occurred. Patterns of resistance included: penicillin only; penicillin and tetracycline; penicillin and chloramphenicol; penicillin, chloramphenicol and tetracycline; and penicillin, cephalothin, erythromycin, clindamycin, chloramphenicol and tetracycline. Many strains also had high-level streptomycin resistance and resistance to co-trimoxazole. The last group of multiply resistant strains occurred mainly in hospitalized children with measles and pneumonia most of whom had received multiple antibiotics. It is probable nosocomial spread of these strains occurred. Strains of *S. pneumoniae* resistant to penicillin, methicillin, carbenicillin, chloramphenicol, streptomycin and cephalothin caused three fatal cases of meningitis in Durban, South Africa.

Strains of *S. pneumoniae* also show low-level resistance to aminoglycosides due to the limited electron transport carried out by these organisms. Effective aminoglycoside entry into bacteria requires terminal electron transport (see Chapter 5). Some strains have also shown high level streptomycin resistance (MIC>2000 μg/ml). This resistance is presumably enzymatic or ribosomal in origin.

Resistance to chloramphenicol has been shown to be from chloramphenicol acetyltransferase in pneumococci. Strains resistant to β-lactams have altered patterns of penicillin-binding proteins determined by SDS polyacrylamide gel electrophoresis and fluorography compared to those of sensitive pneumococci. No β-lactamase has been detected in any strains resistant to β-lactam antibiotics to date.

Genes specifying tetracycline resistance and chloramphenicol acetyltransferase are contained in adjacent heterologous insertions into the chromosome of strains from Japan and Paris. No plasmids have been detected to date. The transfer of antibiotic resistance previously reported is probably transformation of chromosomal genes. Interestingly the presence of the gene for tetracycline resistance enhanced cellular acquisition of the chloramphenicol acetyltransferase gene thus mimicking R-factor mediated transfer.

In a survey of 8 654 isolates of *S. pneumoniae* from Alberta, Canada, Dixon and Lipinski reported 2.2% were eight-fold or greater more resistant to penicillin than control susceptible strains. These authors also reported in other surveys that 0.7% of strains were resistant to erythromycin and 2% to tetracycline. Tetracycline resistance has been reported as high as 23%.

Relatively few serotypes have been reported as antibiotic resistance including types 4, 6, 9, 14, 19 and 23.

Streptococcus pyogenes

Significant resistance to benzylpenicillin has not been reported despite heavy use of the drug for treatment of infections due to *S. pyogenes*. Thus that antibiotic remains the agent of choice for *S. pyogenes* infections.

Tetracycline resistance has been reported frequently enough that this agent should not be currently considered for treatment of *S. pyogenes* infections. Erythromycin and lincomycin resistance in various combinations have been increasingly reported in many parts of the world. Acquired chloramphenicol and sulfonamide resistance also occur commonly. Like all streptococci, low-level resistance to aminoglycosides is a species characteristic.

Plasmids have been detected in *S. pyogenes* and some are transducible by exogenous or endogenous phage. One plasmid is about 17×10^6 in size. Most work has been done with transmissible macrolide-lincosamide resistance. At least some of these strains show erythromycin-inducible resistance to erythromycin and lincomycin and exhibit zonal resistance to lincomycin (i.e. grow in low and high drug concentrations but not at intermediate levels) (see also Chapter 3).

Group D Streptococci

Isolates of *S. fecalis* and other group D streptococci of enteric origin (enterococci) usually exhibit low-level resistance to most antibiotics. Treatment of serious disease due to *S. fecalis* is most often with penicillin or ampicillin plus an aminoglycoside. These organisms possess the low level resistance to aminoglycosides characteristic of most streptococci due to their limited electron transport and, thus, poor entry of aminoglycosides into cells. Most of the group D streptococci also have low-level resistance to penicillin, the basis of which is not understood. It is important to note that *S. bovis,* a group D streptococcus of nonenteric origin, is much more susceptible to penicillin than *S. fecalis*.

The combination of a penicillin and an aminoglycoside is normally synergistic for enterococci. However Moellering and colleagues have shown that some isolates do not show synergism due to high-level resistance to the selected aminoglycoside. These strains have been studied and shown to possess ribosomal mutations producing streptomycin resistance or to have an enzyme modifying the aminoglycoside.

Transmissible resistance by apparent conjugation has been described in *S. fecalis*. Plasmids specifying antibiotic resistance as well as other bacterial properties (hemolysis, bacteriocin) occur. Plasmids vary in size from 6×10^6 to 51×10^6 and may be transmissible or non-transmissible. There is frequently more than one type of plasmid in a single isolate. Resistances transferred include streptomycin, kanamycin, neomycin, gentamicin, tobramycin, sisomicin, netilmicin, erythromycin, lincomycin, tetracycline and chloramphenicol resistance. Apparently, chromosomal determinants (e.g. rifampicin resistance) can also be mobilized by some transmissible plasmids.

In view of the recognized failure of some strains to show antibiotic synergism and the presence of transmissible aminoglycoside resistance it is important to test isolates from serious group D streptococcal disease particularly from endocarditis for antibiotic susceptibility and synergism.

Other Streptococci

In general most other streptococci including anaerobic streptococci pose little in the way of an antibiotic resistance problem except the genus characteristic of low-level aminoglycoside resistance. Group B streptococcal infections of newborn children do not always respond to therapy with penicillin in spite of their susceptibility by laboratory testing. Failure, recurrence and slow response of group B streptococcal sepsis have been documented in response to penicillin by several investigators. The reasons for this involves many factors (see Chapter 1) but one possible reason proposed has been delayed penicillin killing in some strains. However, the *in vivo* significance of this observation remains unclear.

The 'viridans' group of streptococci including *S. mutans, S. mitis, S. sanguis* types I and II, *S. intermedius*-MG and *S. anginosus-constellatus* are generally susceptible to penicillin. Strains of *S. mitis* and *S. sanguis* type II generally have higher MBCs than other strains of the group. Cephalothin and cefamandole were less active than penicillins, and vancomycin and rifampicin were effective agents for this group.

Hemophilus influenzae

In recent years the emergence of ampicillin resistant isolates of *H. influenzae* has occurred in many parts of the world.

Recently a Canadian survey showed that 13% of type b isolates from systemic disease (mainly meningitis and septicemia) were resistant to ampicillin in 1978. This represented an increase from about 7% in 1976. Resistance was found in all regions of the nation. Ampicillin resistance has also been detected in non-encapsulated strains which are important etiological agents of otitis media in young children.

Resistance to other antibiotics is also now frequently recognized. Tetracycline resistant strains are readily detected but have not exceeded 10% of strains in most surveys. Of particular concern, chloramphenicol resistant strains have been occasionally recognized. Strains resistant to combinations of ampicillin and other antibiotics are also detected. Transmissible resistances to ampicillin, ampicillin and tetracycline, tetracycline and chloramphenicol and on one occasion to ampicillin, tetracycline and chloramphenicol have been reported. Resistance to kanamycin and other aminoglycosides (not used for *H. influenzae* infections) is also known to occur. Resistance to cotrimoxazole remains uncommon.

Ampicillin resistance of *H. influenzae* is mediated in most cases by TEM type β-lactamases. Two general size classes of plasmid DNA occur among strains of *H. influenzae*. The smaller group of molecular weight 2.5×10^6–4.4×10^6 are non-transmissible and contain a portion of a transposable element (TnA) which specified a TEM β-lactamase. The large $30–40 \times 10^6$ molecular weight plasmids are frequently self-transmissible by conjugation and may contain resistance genes for tetracycline, chloramphenicol and kanamycin as well as ampicillin. They possess the entire TnA sequence. Several large plasmids from diverse geographic locations show considerable DNA homology (60% or greater base sequence homology) and have similar restriction endonuclease digestion patterns on agarose gel electrophoresis. Their guanosine and cytosine base composition of about 39% is similar to that of the *H. influenzae* chromosome. These observations are compatible with the view that the transmissible plasmids are resident in *H. influenzae* or a closely related species and did not originate in enteric bacteria.

Evidence has also been advanced supporting the presence of

antibiotic resistances in the chromosome including resistance to tetracycline, chloramphenicol and ampicillin. Chromosomal resistance was considered the basis of resistance in several strains examined by Stuy. It appears that conjugative plasmids integrate into the *H. influenzae* chromosome. These do not conjugally transfer resistance unless excised. Thus, plasmid minus strains with antibiotic resistance probably have plasmids but they are chromosomally integrated.

Transfer of plasmids has been generally reported to be more effective on solid medium or in pellets following centrifugation and is apparently more efficient between identical serotypes.

Several of the small plasmids of *H. influenzae* and *N. gonorrhoeae* have considerable shared DNA sequence homology (70% or greater in several studies). A 4.4×10^6 molecular weight plasmid has been shown to contain about 40% of the TnA commonly found in R-factors of enteric origin but does not contain both inverted repeats. These plasmids do not show significant base homology with the large plasmids of *H. influenzae* or *N. gonorrhoeae*. From studies of small plasmids of *H. influenzae* and *N. gonorrhoeae* it is clear there is no relationship between large and small plasmids. This raises the possibility that the small plasmids had a common origin which could be enteric or other bacteria.

Many strains of *H. parainfluenzae* have β-lactamase activity and are generally more antibiotic resistant than strains of *H. influenzae*.

Several investigators have examined the susceptibility of β-lactamase producing strains of *H. influenzae* to other β-lactams, particularly new cephalosporins. The most active of those agents seem to be cefotaxime and moxalactam. Slightly less active are cefoperazone, cefoxitin and cefmetazole. Also very active on most strains are cefamandole, cefuroxime and cefotiam. Oral cephalosporins such as cefaclor, cephaglycin and cefatrizine are considerably less active on β-lactamase minus strains.

Strains producing β-lactamase have a reduced rate of lysis and show persistence of spherical forms when treated with cefamandole and these can form colonies. The meaning of these observations to *in vivo* susceptibility is not yet clear but implies a possible basis of persistence of *H. influenzae* strains causing meningitis.

MICs of other *β*-lactams including carbenicillin, ticarcillin, piperacillin are elevated for *β*-lactamase positive strains. MICs may, on occasion, be many times that for *β*-lactamase minus strains.

Neisseria spp.

Neisseria gonorrhoeae

The most widely used antibiotic for therapy of gonorrhoea is benzyl penicillin. Beginning in the middle 1950s, increasing numbers of therapeutic failures with penicillin were observed. At that time and subsequently, a number of strains have been isolated which are less susceptible to penicillin. These strains exhibited only low-level resistance and rarely had MICs of penicillin in excess of 2 μg/ml. The majority of gonococcal genital infections due to these strains could be still effectively treated with 4.8 million units of penicillin used with 1 g of oral probenecid. Ampicillin in doses of 3 to 4 g with probenecid was also generally effective. With the introduction of these massive doses of penicillin, the frequency of low-level resistance to penicillin has generally ceased to increase and in some communities has definitely decreased.

Strains of gonococci with low-level penicillin resistance frequently exhibit some increase in resistance to other antibiotics including tetracycline, chloramphenicol, erythromycin, rifampicin and fusidic acid. High-level resistance to streptomycin may also be associated.

Three genetic loci are known to affect resistance to several drugs, dyes and detergents. The *mtr* and *pen*B loci are associated with increased resistance whereas the *env* locus is associated with increased susceptibility. The *mtr*-2 locus is associated with an increase of a 52 000 molecular weight outer membrane protein and has less phospholipid in the outer membrane. This is associated with reduced susceptibility to hydrophobic antibiotics. Changes in an outer membrane protein previously described associated with the *pen*B locus are due to another mutation. The changes associated with the *mtr*-2 mutation suggests that such resistance is due to reduced drug permeability. Preliminary studies on the *env* mutations show it results in an increase of phospholipid in the outer membrane and thus, presumably less polysaccharide. The barrier to hydrophobic agents is thus reduced.

The *pen*A locus in contrast to the preceding mutations causes increased resistance to penicillin only. Combinations of the resistant mutations can cause significantly higher levels of resistance. For example *pen*A2 mutants are about 8-fold more resistant to penicillin whereas *pen*A2, *mtr*-2 mutants are 26-fold more resistant.

It has also been shown that membranes of strains which are penicillin resistant but β-lactamase negative bind less ^{14}C-penicillin G than do membranes of sensitive strains. Thus penicillin-binding proteins could be altered in membranes of low-level resistant strains.

Resistance to some drugs particularly high-level streptomycin resistance is due to a single-step ribosomal mutation (*str*). Spectinomycin resistance is very uncommon among clinical isolates but has been shown to be a ribosomal mutation (*spc*). Many other genetic loci have been shown to produce antibiotic resistance including 2- to 4-fold increases for chloramphenicol (*cam*), tetracycline (*tet*) and erythromycin (*ery*) and 100-fold or greater increase for rifampicin (*rif*) and fusidic acid (*fus*).

In 1976 isolates of *N. gonorrhoeae* possessing β-lactamase and resistant to penicillin and ampicillin appeared in the Far East and England. These isolates now occur in many areas of the world although the frequency of isolation varies considerably. These have caused genital infections but also salpingitis and disseminated infections as well as other infections.

The β-lactamases described to date in gonococci are TEM-1 like enzymes. These enzymes can be inhibited by sodium clavulanate resulting in increased susceptibility to penicillin and ampicillin.

Plasmids associated with β-lactamase production carry about 40% of Tn2 and are closely related to some small plasmids of *H. influenzae*. Gonococcal isolates from the Far East were found to contain a 4.4×10^6 molecular weight plasmid and those from London and Liverpool England had a 3.2×10^6 plasmid. These plasmids and a *H. influenzae* plasmid are closely related by DNA sequence homology and have a 0.4 to 0.41 guanosine plus cytosine mole fraction. The plasmids are nonconjugative.

A 24.5×10^6 plasmid has been detected in strains of *N. gonorrhoeae* capable of transferring antibiotic resistance by conjugation. The same plasmid is occasionally found in penicillin-sensitive strains. It does not apparently carry antibiotic resistance. The guanosine plus cytosine mole fraction of this plasmid is similar to that of the chromosome of *N. gonorrhoeae* (0.50). The 24.5×10^6 plasmid is needed for transformation of chromosomal genes into non-piliated recipient cells.

It has been reported that strains producing β-lactamase also show some elevation of MICs of tetracycline and erythromycin, and to a minor degree, of spectinomycin.

Several agents are active on β-lactamase positive strains. These include rosaramicin, several new cephalosporins including cefuroxime, cefaclor, cefoxitin, cefamandole, and, for many strains, tetracycline, spectinomycin and kanamycin. As noted β-lactamase inhibitors such as clavulanic acid are effective at inhibiting the currently prevalent β-lactamase and at reducing MICs of penicillin and ampicillin.

Strains producing disseminated gonococcal infections rarely show low-level resistance to penicillin. Generally such strains isolated from diverse areas have been highly susceptible to penicillin and frequently show similar patterns of auxotrophy. Strains with β-lactamase can cause disseminated infections but it remains to be seen if they will do so commonly.

Neisseria meningitidis

Strains of *N. meningitidis* remain almost entirely susceptible to the therapeutic agents of choice, benzylpenicillin, ampicillin and chloramphenicol. Extensive resistance to sulfonamides in several sero-groups has been described. The criteria for definition of resistance to sulfonamides have varied. Those of Feldman are: susceptible – inhibited by 1 μg/ml or less; resistant – not inhibited by 10 μg/ml; 'partially-resistant' – not inhibited by 1 μg/ml but inhibited by 10 μg/ml.

Resistance to rifampicin has emerged readily during courses of prophylaxis for meningococcal carriage although this agent is currently recommended for such purposes. Minocycline prophylaxis has not been associated with resistance but often produces disturbing vertigo when used.

Shigella

Multiply resistant strains of *Shigella* in Japan lead to the first detection of transferable drug resistance opening the door to a vast number of subsequent studies of R-factors. In general, antibiotic resistance remains widespread among isolates of *Shigella* in many areas of the world. Multiple resistance is frequent and is due almost always to R-factors. Resistance to tetracyclines, streptomycin, sulfonamides and ampicillin have been widely reported. However, chloramphenicol resistance has been generally uncommon in the US and Europe although frequent in Japan, Central America and Mexico. Resistance to co-trimoxazole and to quinoline drugs including nalidixic acid has been reported to be very low.

If antibiotics are to be used for therapy of Shigellosis, local antibiotic susceptibility patterns should be known or the results of susceptibility testing awaited. Many less severe cases of Shigellosis do not require antibiotic therapy.

Most experience has indicated that once ampicillin resistance is frequent among strains of *Shigella,* it remains predominant. However occasionally the acquisition of ampicillin and other resistances and subsequent loss of these in a specific geographic area has been reported.

Antibiotic-resistant *Shigella dysenteriae* type 1 have caused several epidemics of bacillary dysentery. The resistance is plasmid mediated, most often by plasmids of Inc group B. Epidemics in Mexico and Central America were associated with strains resistant to chloramphenicol, tetracycline, streptomycin and sulfonamides. Strains from subsequent outbreaks in Mexico, Central America and Bangladesh were also ampicillin resistant. Strains from the different epidemics had a variety of plasmids including a 5.5×10^6 probably identical plasmid which contained TnA. This small nonconjugative plasmid was also detected in other enterobacteriaceae isolated. Ampicillin-resistant strains of *S. typhi* isolated from typhoid epidemics in Mexico had plasmid DNA that had some homology with the 5.5×10^6 plasmids and probably contained a portion of TnA. TnA is thus distributed, in part or completely, among *Salmonella* and *Shigella* capable of severe, potentially life-threatening infections from different geographic locales.

Salmonella

Resistance to various antimicrobial agents among food-poisoning strains of *Salmonella* and *Salmonella typhi* varies with time and geographic location. Transferable resistance has been found in both groups of organisms. Resistance to streptomycin, sulfonamides and tetracycline, in particular, has been common. Resistance to ampicillin and to chloramphenicol has generally occurred at a lower frequency but the incidence has varied significantly in different parts of the world. Co-trimoxazole resistance is uncommon to date.

Antibiotic resistance among food-poisoning strains of *Salmonella* and particularly *Salmonella typhimurium* has been shown to develop in calves and spread to humans. Studies in Great Britain have shown antibiotic resistance emerged among calves fed tetracycline. Subsequently, such resistant strains could be shown to spread to dairy cattle and to humans. Although calves fed antibiotics develop resistant *S. typhimurium,* resistance can also apparently develop in response to therapeutic use of antibiotics. Strains of *S. typhimurium* phage type 204 carrying an H2 plasmid specifying resistance to chloramphenicol, streptomycin, sulfonamides and tetracyclines have appeared in calves treated with chloramphenicol. Some isolates of phage type 204 have appeared among human cases in Britain.

Work in North America has also followed the development of chloramphenicol resistance in animals. It has been concluded that this chloramphenicol resistance was not due to supplementation of animal feed with antibiotics but rather due to treatment of animals with chloramphenicol. In these studies, plasmids found in *Salmonella typhimurium* from both humans and animals were both IncH and non-IncH type R-factors. Chloramphenicol was mainly carried by H-type R-factors whereas ampicillin resistance was not specified by the heat-sensitive transfer system seen with H-plasmids.

Not surprisingly antibiotic resistance in *S. typhimurium* can develop independently in humans undergoing therapeutic use of antibiotics. Ampicillin resistance, for example, developed in humans in the New York area prior to its detection in animals.

In Japan, it has been reported that resistance to four antibiotics among strains of *Salmonella* was uncommon (2.2%). Single resistance to sulfonamides was very common. Much less common, but considerably more common than quadruple resistance, was triple resistance to tetracyclines, streptomycin and sulfonamides. R-factors were present and most common in strains showing multiple resistance.

Food-poisoning infections in the vast majority of cases do not require treatment with antibiotics. However, some of these strains of *Salmonella* may produce tissue or blood-borne infections. It has been shown that although chloramphenicol resistance is uncommon in North America, treatment of such infections with chloramphenicol has been associated with the development of chloramphenicol resistance during treatment. In most of these strains chloramphenicol resistance was unstable.

A major concern has been the development of chloramphenicol resistance among strains of *Salmonella typhi* responsible for large numbers of cases particularly in Mexico but also in India, Vietnam, Thailand and Spain. Chloramphenicol-resistant strains of *Salmonella typhi* have been very uncommon in most of Europe and in North America outside of Mexico. It has been reported that more than 10 000 cases of typhoid were produced in 1972 and 1973 in Mexico by chloramphenicol-resistant *S. typhi* strains. The plasmid in these strains was an H-1 plasmid and also specified resistance to sulfonamides, tetracycline and streptomycin. Some chloramphenicol resistant strains have also carried resistance to ampicillin. Ampicillin may also be carried on an H-1 plasmid but has been reported also on an I group plasmid which does not have a heat-sensitive transfer system.

Undoubtedly the situation is even more complex than indicated above. However, it is clear that the capability for chloramphenicol resistance among *Salmonella typhi* strains has been achieved in many parts of the world. It should be noted that the incidence of chloramphenicol resistance has subsequently declined. In some instances strains have been ampicillin resistant either with or without chloramphenicol resistance. A significant number of strains resistant to trimethoprim-sulfamethoxazole have not been reported.

Gastrointestinal pathogens other than Shigella and Salmonella

Campylobacter jejuni/coli

C. jejuni is a frequent cause of diarrhea which is often bloody and associated with abdominal pain. It is particularly important in children and has been reported second to food-poisoning *Salmonella* as the most frequent cause of diarrhea in that group. Ileocolitis can also be produced.

Antibiotic treatment is not indicated for most cases of Campylobacter enteritis. Erythromycin has been widely proposed as the drug of choice when drug therapy is necessary. In some cases significant resistance to erythromycin has been reported. Such resistance has included lincomycin and clindamycin and was 8% of strains in one study. Gentamicin is active with little or no resistance detected. For serious infections gentamicin requires parenteral use as its efficacy orally has not been established. Resistance to tetracycline and doxycycline is very low. The nitrofuran, furazolidone is also very active.

C. jejuni is resistant to most β-lactam antibiotics including new cephalosporins such as cefuroxime and cefotaxime. Strains produce a β-lactamase active on penicillins. Resistance is extensive to vancomycin and rifampicin.

Yersinia enterocolitics

Y. enterocolitica can produce a variety of clinical syndromes with enterocolitis the most common of these. All strains produce a heat stable toxin and some are invasive for gut epithelial cells. In general, aminoglycosides including gentamicin, tobramycin, amikacin and kanamycin are the most active group of antibiotics on these organisms which only occasionally require antibiotic therapy. Tetracyclines and chloramphenicol are also highly active although occasional resistance has been reported. In general, strains are resistant to β-lactam antibiotics and produce a β-lactamase. Unlike *C. jejuni*, *Y. enterocolitica* has very low susceptibility to erythromycin and clindamycin.

Vibrio cholerae

Plasmid-specified antibiotic resistance has been detected sporadically among isolates of *V. cholerae*. Strains isolated from Asia were shown to contain plasmids of the IncC group which specified resistance to tetracycline, streptomycin, chloramphenicol, sulfonamides and kanamycin. Non-transferable as well as transferable plasmids were detected.

Recently a widespread outbreak of R-factor specified antibiotic resistance among El Tor *V. cholerae* in Tanzania has been detected. In association with heavy usage of prophylactic tetracycline, 76% of strains were resistant to tetracycline 6 months after the epidemic was recognized. Plasmids were IncC group and resistances transferred included tetracycline, ampicillin, sulfonamide, chloramphenicol, kanamycin and streptomycin.

Enterotoxigenic E. coli

The antibiotic susceptibility patterns of *E. coli* strains producing heat labile (LT) or heat stable (ST) or both toxins have varied from different geographic locations. In Kenya antibiotic resistance was not significant in toxigenic *E. coli* although it was in non-toxigenic strains. Even after the use of prophylactic doxycycline in volunteers resistance did not appear in toxigenic strains although it increased in non-toxigenic strains. Unquestionably doxycycline provoked resistance in the latter strains but the resistance frequency declined to levels similar to those of volunteers not receiving doxycycline after prophylaxis had been stopped for two weeks.

In sharp contrast toxigenic strains isolated from the Far East showed a high degree of multiple antibiotic resistance. Resistance was frequently transferable and associated with toxin-producing capability in several cases. This situation raised the possibility that antibiotic usage could select for plasmids specifying toxin production and antibiotic resistance. At least some plasmids specifying ST and LT toxin are carried by IncF1 plasmids. Resistance to ampicillin, chloramphenicol, streptomycin, sulfonamides, tetracycline and cephaloridine was most common.

The preceding results show that antibiotic resistance does occur in toxigenic strains. If antibiotic prophylaxis is to be considered, the risk of provoking enhanced levels of resistance exists. However, short-term use in visitors may be acceptable because resistance is likely to be unstable in the absence of antibiotic use.

Nosocomial gram-negative bacteria

Enterobacteriaceae

Almost every major resistance mechanism has been described in *E. coli* or other Enterobacteriaceae. However, the prevalence and significance of resistance varies with time, location and type of infection. Nosocomial outbreaks of infection with multiply resistant strains have been a major problem with many of these bacteria. This has been particularly so with *Klebsiella, Serratia, Proteus, Providencia,* and *Enterobacter* isolates.

Nosocomial outbreaks often involve spread of a single multiply resistant strain through a unit or hospital. However, outbreaks can involve more than a single strain or species and involve both strain and plasmid spread.

In general, emergence of nosocomial resistance problems among Enterobacteriaceae requires a resistant strain usually due to resistance plasmids, selective pressure of antibiotic use, a strain capable of colonization and some virulence and a group of susceptible patients. Situations where these factors may be operative include intensive care, urology, burn and neurosurgical units.

Klebsiella pneumoniae is an important cause of nosocomial respiratory tract and urinary tract infections and bacteremias. In many instances multiply resistant strains possessing plasmids produce these infections following initial intestinal colonization. However, index cases initiating outbreaks may be individuals admitted to hospital with overt infection.

Outbreaks of gentamicin-resistant *Klebsiella* have become relatively common. Such resistance is usually due to R-factor specified enzymes although non-enzymatic aminoglycoside resistance has also been reported. Resistance to other aminoglycosides, penicillins, cephalosporins, chloramphenicol and sulfonamides has been frequently transferred.

Serratia marcescens is frequently involved in similar nosocomial outbreaks to *Klebsiella*. Studies have now shown that *S. marcescens, Klebsiella* species and *Enterobacter aerogenes* isolates from one such outbreak possessed an identical R-plasmid specifying aminoglycoside and β-lactam resistance which had spread among the strains. Other investigations have shown that such 'plasmid-epidemics' can involve various Enterobacteriaceae including *E. coli*.

A striking illustration of the selective role that antibiotic use can play has been illustrated by the control of *K. aerogenes* infections in a neurosurgical unit by withdrawl of all antibiotics.

In addition to the organisms listed earlier, other antibiotic-resistant Enterobacteriaceae and non-Enterobacteriaceae such as *Citrobacter, Morganella, Acinetobacter* and others can cause both nosocomial outbreaks as well as isolated endemic infections in patients with compromised host defenses. Urinary infections are particularly common especially in male patients with urinary catheters. A second major type of such infections is respiratory infections in intensive care units.

These organisms represent a mixture of intrinsic resistance and plasmid-mediated resistance. The former is probably of greater importance in bacteria like *Providencia* and *Acinetobacter*. Plasmids may occur in any but are of marked importance in *Klebsiella, Serratia* and *E. coli*. Multiply resistant organisms are selected particularly by the use of broad spectrum agents such as gentamicin, cephalosporins and co-trimoxazole.

The Enterobacteriaceae pose a much smaller resistance problem among patients who are not in hospital. *E. coli* is a major cause of urinary tract infections. Nalidixic acid resistance emerging during a course of treatment of these infections is not uncommon. Resistance to sulfonamides and ampicillin occurs in varying degrees at different locations. In general, among community-based patients, it has not posed a major problem. The same has been true for co-trimoxazole even in situations of long-term use. However, resistance to co-trimoxazole, while uncommon under these circumstances, has been documented.

Several of the Enterobacteriaceae are intrinsically resistant to drugs frequently active on gram-negative bacteria. For example, *Enterobacter* spp. are often resistant to cefazolin and cephalothin in part due to a cephalosporinase; *Klebsiella pneumoniae* often show low-level ampicillin resistance; *Proteus* spp. are resistant to polymyxin; and *Providencia* are resistant to cephalothin. Susceptibility to many agents cannot be predicted and testing is needed in any serious sepsis due to this group of bacteria.

Pseudomonas aeruginosa

P. aeruginosa remains an important cause of nosocomial infections. The organism has a high degree of intrinsic resistance to many antibiotics. The most active agents are tobramycin, sisomicin, piperacillin, cefsulodin and azlocillin. Other antibiotics have reasonable activity but less than the above. These include: gentamicin, amikacin, cefotaxime, cefoperazane and moxalactam. Susceptibility to aminoglycosides is highly dependent on the ionized divalent cation concentrations in the testing medium, particularly calcium and magnesium.

Acquired additional resistance superimposed on intrinsic resistance has also been a problem. R-factor mediated resistance occurs particularly associated with topical antibiotic use, burn units and urinary tract infections. R-plasmids fall into two major groups: broad and narrow host range. The narrow-range plasmids particularly Inc group P-2 are the most prevalent but wide host range plasmids of Inc groups P-1 and P-3 (C) also have been detected. Overall, R-factors belonging to more than 11 groups have been detected.

Resistances to antipseudomonal penicillins and aminoglycosides as well as to tetracycline, sulfonamides, chloramphenicol, mercury, boron, tellurium and miscellaneous other agents are found on plasmids. Plasmid-specified TEM-1, TEM-2, PSE-1, PSE-2, PSE-3, PSE-4, OXA-2, OXA-3 and SHV-1 β-lactamases have been detected. Many of the new cephalosporins are either poorly or not hydrolyzed by these enzymes. Cefsulodin and cefoperazone are relatively resistant and cefotaxime and moxalactam highly resistant to the β-lactamases. Various aminoglycoside acetylating, nucleotidylating and phosphorylating enzymes occur on plasmids.

Non-enzymic plasmid-specified resistance to kanamycin and streptomycin has also been described, although uncommonly.

A common form of aminoglycoside resistance is found in many strains under various circumstances in most major hospitals. These strains show some resistance to all aminoglycosides. The level of resistance varies from susceptible to highly resistant but most strains have intermediate resistance levels. They have susceptible targets, no detectable enzymes and may or may not have plasmids. Resistance transfer by conjugation has not been reported. These strains accumulate less aminoglycoside than do susceptible strains at similar concentrations. Resistance is due to an apparent reduction of permeability to aminoglycosides as a group and may be associated with a change in outer membrane proteins.

The permeability form of resistance most often emerges during a course of aminoglycoside treatment and is best controlled by limiting use of all aminoglycosides. These strains can also spread between patients in hospital wards.

A major and often poorly recognized fact is that many strains of *P. aeruginosa* act resistant *in vivo* to a drug they have been tested susceptible to in the laboratory. This is partly explained in the case of aminoglycosides by tissue antagonism of the drugs and often poor entry into some tissues (see Chapter 1). Several examples have been reported where response to treatment of septicemias and bronchopneumonias did not correlate with susceptibility status. This is an important point because the laboratory results tend to overestimate susceptibility and encourage inappropriate dosing, inadequate attention to dosage regimens, blood and tissue levels and a failure to select the most active of the available agents. Introduction of potent penicillins like piperacillin and cephalosporins like cefsulodin may alleviate this problem in the future.

Other types of *Pseudomonas* also cause nosocomial infections on occasion. These include *P. cepacia* and *P. maltophilia* both of which are intrinsically resistant to many antibiotics. *P. cepacia* produces at least one type of β-lactamase and plasmid DNA has been detected in some strains. The polymyxin resistance seen in this strain is apparently specified by the cytoplasmic membrane rather than the outer membrane.

Legionella pneumophila and other similar agents

The antibiotic of choice for *L. pneumophila*, Pittsburg Pneumonia Agent (*L. bozemanii*) the Wiga agent (*L. Mc. dadeii*) and Tex KL agents is erythromycin with or without rifampicin. Tetracyclines and co-trimoxazole are preferred alternative agents. Gentamicin, tobramycin, amikacin, cefoxitin, chloramphenicol and minocycline are active by *in vitro* testing but their *in vivo* effectiveness is not established.

Several serotypes of *L. pneumophila* possess a β-lactamase which is principally a cephalosporinase but also hydrolyzes penicillins. The β-lactamase inhibitor CP-45,899 effectively prevents inactivation of cefamandole, clavulanic acid less so. Cefoxitin and cefuroxime are not hydrolyzed by the enzyme.

Anaerobic bacteria

The anaerobic gram-negative bacillus *Bacteroides fragilis* is an important and frequently isolated anaerobe from clinical specimens. *B. fragilis* including most isolates of the various subspecies of this group are relatively resistant to penicillins and to β-lactamase – susceptible cephalosporins. This resistance is due, in part, to the widespread possession of β-lactamases by these bacteria. These β-lactamases are principally cephalosporinases (see Chapter 3). It seems likely that poor penetration of the cell envelope also contributes to some extent to this form of resistance.

Cephalosporins with reduced susceptibility to β-lactamases show increased activity against the *B. fragilis* group. Particularly active are cefoxitin, cefmetazole and moxalactam. Thienamycin is also very active on most *B. fragilis*. Cefoxitin-resistant strains are uncommonly isolated and show resistance to many of the β-lactamase-resistant cefamycins and to thienamycin. Some of these strains hydrolyze thienamycin but not cefamycins nor the β-lactamase inhibitor, CP-45,899. The K_m values of the enzymes for thienamycin have not been reported. This parameter is probably of greater importance in determining the role of the β-lactamase in mediating thienamycin resistance than simply knowing that hydrolysis occurs.

Tetracycline resistance is widespread among *B. fragilis*. Resistance is mediated by plasmids which may be inducible both for tetracycline resistance and transferability. Such induction has also been obtained *in vivo* by feeding mice tetracycline. Constitutive mutants expressing tetracycline resistance are readily obtained *in vitro* and are found *in vivo* when mice are fed high doses of tetracycline. The mechanism of resistance is described in Chapter 3.

Resistance to clindamycin and erythromycin is currently uncommon in *B. fragilis* isolates. However, it has been detected and is transferable. In some instances transfer depended on induction of a tetracycline-transfer system whereas in another case it was associated with 20×10^6 and 2×10^6 molecular weight plasmids. Tetracycline resistance which is apparently non-plasmid linked has recently been described. Chloramphenicol resistance remains very uncommon. Considerable variation of susceptibility of *B. fragilis* particularly to clindamycin but also to chloramphenicol, tetracycline and doxycycline has been reported. This is seen between different geographical areas but also has been reported from hospitals in the same city.

Metronidazole is a very effective agent for *B. fragilis* in that resistance is uncommon and it penetrates many tissues, including brain, well. Other 5-nitroimidazoles are also active including ornidazole and SC-28,538. A few isolates resistant to metronidazole have been found. One of these has been shown to have decreased levels of pyruvate dehydrogenase activity. This could result in a reduced rate of metronidazole reduction (see Chapter 2).

Anaerobic bacteria are resistant to aminoglycoside antibiotics because of an absence of oxygen or nitrate dependent terminal electron transport that results in a failure of transport of aminoglycosides.

Most other clinically important anaerobic bacteria including other species of *Bacteroides* have retained susceptibility to benzylpenicillin and other penicillins. Rare isolates of Clostridia have been found resistant to β-lactam antibiotics. Some of these have produced β-lactamases of broad spectrum type and were inducible to variable extents. Plasmid-specified β-lactamase has been demonstrated in Clostridia.

Clostridium difficile is the major causative organism of antibiotic-associated diarrhea. Severe forms of such diarrhea have been seen following clindamycin and other antibiotic use, particularly ampicillin. *C. difficile* strains isolated may be relatively sensitive to clindamycin although strains with lower susceptibility are encountered. A recent study showed 60% were susceptible to 1 μg/ml or less but 9% required 128 μg/ml or more. It seems that the process is not simply a superinfection with clindamycin- or ampicillin-resistant *C. difficile*.

Strains are susceptible to fecal levels of vancomycin but can persist as spores. Germinated spores can be responsible for a recurrence of the illness.

Fusobacteria and Peptostreptococci are almost always susceptible to benzylpenicillin, chloramphenicol and clindamycin. The new cephalosporins including moxalactam, cefotaxime, cefoxitin, cefamandole and cefuroxime are very active on many anaerobic gram-positive cocci and uncommonly isolated anaerobes. However, benzylpenicillin is frequently at least as active as these cephalosporins and also many new semisynthetic penicillins.

Mycobacterium tuberculosis

Drug resistance to the various antituberculous agents in previously untreated patients (primary-resistance) is usually low, being only a few per cent at most. However, much higher primary resistance rates for these agents have been reported particularly from several areas including the United States. In general, surveillance susceptibility testing of isolates from previously untreated patients should be carried out to provide the community pattern of susceptibility. Communities with a substantial number of cases in people immigrating into the area, should also carry out such surveillance.

Resistance to agents can account for treatment failure. The use of multiple drugs in treatment has helped reduce the effects of resistant strains in this respect. To date most resistance has been mutational so that a strain resistant to rifampicin for example is very likely susceptible to isoniazid.

Susceptibility testing should be performed for isolates from previously treated patients, treatment failures during treatment including lack of response and reversion of sputum to culture positive and patients from areas of the world with high or unknown levels of drug resistance.

Resistance to a drug by *in vitro* testing does not necessarily mean that a drug will fail. An explanation for this observation is that resistance may be overcome by the use of multiple antibiotics. The atypical bacteria are frequently resistant to antituberculous drugs but may respond to their clinical use.

Miscellaneous bacteria

Mycoplasma and Ureaplasma are resistant to agents acting on peptidoglycan synthesis. They are frequently susceptible to tetracycline and erythromycin. However, strains resistant to 30 times or more the concentration of tetracycline needed to inhibit susceptible strains of *U. urealyticum* have been found in over 9% of isolates in one study. These remained susceptible to erythromycin.

Chlamydia trachomatis susceptibility to antibiotics can be tested by using cell culture methods. These organisms are usually treated with tetracyclines, erythromycin or chloramphenicol. Rifampicin is very active as is rosaramicin. Chlorhexidine and povodone-iodine are active as disinfectants. No significant resistance to anti-chlamydial agents has been reported.

Tetracycline with or without streptomycin is recommended for therapy of Brucellosis. Relapses are rarely due to acquired resistance to tetracyclines. Many agents are of poor effectiveness for Brucellosis perhaps because they enter the phagocytic vacuoles containing *Brucella* organisms poorly. Co-trimoxazole may also be effective.

Preferred treatment of tularemia is with streptomycin. Chloramphenicol and tetracyclines usually fail to eradicate *Francisella tularensis*. Relapses, like those with *Brucella,* are rarely due to drug resistance as the patient will often respond to another course of the same antibiotic.

Hemophilus ducreyi can carry resistance plasmids specifying an entire TnA transposon coding for a TEM-1 type β-lactamase and tetracycline resistance. Resistance was carried on 6×10^6 MW nonconjugative plasmid. Strains from an outbreak in Winnipeg Canada were all susceptible to sulfisoxazole, nalidixic acid, rifampicin and chloramphenicol.

Gardnerella (Haemophilus, Corynebacterium) vaginalis is susceptible to penicillin, erythromycin, cefazolin, cephalothin and several other agents. They are relatively resistant to many aminoglycosides, colistin, sulfonamides, trimethoprim and nalidixic acid. Favorable therapeutic responses to metronidazole have been reported.

Antibiotic resistance has not been described as of significance for *Yersinia pestis, Pasteurella multocida* and *Bordetella pertussis.* The antibiotic of choice for *B. pertussis* is erythromycin. The response of *B. pertussis* to erythromycin is often questionable.

Selected references

Applebaum, P.C., Seragg, J.N., Bowen, A.J., Bhamjee, A., Hallett, A.F. and Cooper, R.C. (1977). *Streptococcus pneumoniae* resistant to penicillin and chloramphenicol. *Lancet* **ii,** 995–7.

Atkinson, B.A. (1980). Species incidence, trends of susceptibility to antibiotics in the United States, and minimum inhibitory concentration. *Antibiotics in Laboratory Medicine*, ed. V. Lorian, pp. 607–722. Williams and Wilkins, Baltimore.

Bagg, R. (1978). Antibiotic treatment of Staphylococcal pneumonia in adults. *J. Antimicrob. Chemother.* **4,** 297–8.

Baker, C.N., Thornsberry, C. and Jones, R.N. (1980). *In vitro* antimicrobial activity of cefoperazone, cefotaxime, moxalactam, (LY127935), azlocillin, mezlocillin and other β-lactam antibiotics against *Neisseria gonorrhoeae* and *Haemophilus influenzae,* including β-lactamase-producing strains. *Antimicrob. Agents Chemother.* **17,** 757–61.

Blackman, H.J., Yoneda, C., Dawson, C.R. and Schachter, J. (1977). Antibiotic susceptibility of *Chlamydia trachomatis. Antimicrob. Agent Chemother.* **12,** 673–7.

Bourgault, A.M., Wilson, W.R. and Washington, J.A. (1979). Antimicrobial susceptibilities of species of viridans streptococci. *J. Infect. Dis.* **140,** 316–21.

Britz, M.L. and Wilkinson, R.G. (1979). Isolation and properties of metronidazole-resistant mutants of *Bacteroides fragilis. Antimicrob. Agents Chemother.* **16,** 19–27.

Bryan, L.E. (1976). Gentamicin resistance in clinical-isolates of *Pseudomonas aeruginosa* associated with diminished gentamicin accumulation and no detectable enzymatic modification. *J. Antibiotics.* **29,** 743–53.

Bryan, L.E. (1978). Transferable chloramphenicol and ampicillin resistance in a strain of *Haemophilus influenzae. Antimicrob. Agents Chemother.* **14,** 154–6.

Bryan, L.E. (1979). Resistance to antimicrobial agents: the general nature of the problem and the basis of resistance. In *Pseudomonas aeruginosa Clinical Manifestations of Infection and Current Therapy,* ed. R.G. Doggett pp. 219–70. Academic Press, New York.

Bryan, L.E., Kowand, S.K. and Van Den Elzen, H.M. (1979). Mechanism of aminoglycoside antibiotic resistance in anaerobic bacteria: *Clostridium perfringens* and *Bacteroides fragilis. Antimicrob. Agents Chemother.* **15,** 7–13.

Byers, P.A., Dupont, H.L. and Goldschmidt, M. (1976). Antimicrobial susceptibilities of shigellae isolated in Houston, Texas in 1974. *Antimicrob. Agents Chemother.* **9,** 288–91.

Cherubin, C.E., Neu, H.C., Rahal, J.J. and Sabath, L.D. (1977). Emergence of resistance to chloramphenicol in Salmonella. *J. Infect. Dis.* **135,** 807–12.

Chow, A.W., Taylor, P.R., Yoshikawa, T.T. and Guze, L.B. (1979). A nosocomial outbreak of infections due to multiply resistant *Proteus mirabilis:* role of intestinal colonization as a major resevoir. *J. Infect. Dis.* **139,** 621–7.

Clewell, D.B. and Franke, A.E. (1974). Characterization of a plasmid determining resistance to erythromycin, lincomycin and vernamycin B_{α} in a strain of *Streptococcus pyogenes. Antimicrob. Agents Chemother.* **5,** 534–7.

Crossley, K., Loesch, D., Landesman, B., Mead, K., Chern, M. and Strate, R. (1979). An outbreak of infections caused by strains of *Staphylococcus aureus* resistant to methicillin and aminoglycosides. *J. Infect. Dis.* **139,** 273–9.

Davies, P.A. (1979). Infections with gram-negative rods. *J. Antimicrob. Chemother.* **5** (suppl. A), 13–20.

Del Bene, V.E., Farrar, W.E., Weinrich, A.E., Brunson, J.W. and Rubens, C.E. (1977). *β*-lactamase, *β*-lactam resistance and extrachromosomal DNA in anaerobic bacteria. In *Microbial Drug Resistance,* vol. 2, ed. S. Mitsuhashi, pp. 310–12, University of Tokyo Press, Tokyo.

Dixon, J.M.S., Lipinski, A.E. and Graham, M. (1977). Detection and prevalence of pneumococci with increased resistance to penicillin. *Can. Med. Assoc. J.* **117,** 1159–61.

Dornbush, K., Olsson-Liljequist, B. and Nord, C.E. (1980). Antibacterial activity of new *β*-lactam antibiotics on cefoxitin-resistant strains of *Bacteroides fragilis. J. Antimicrob. Chemother.* **6,** 207–16.

Dzink, J. and Bartlett, J.G. (1980). *In vitro* susceptibility of *Clostridium difficile* isolates from patients with antibiotic-associated diarrhea or colitis. *Antimicrob. Agents Chemother.* **17,** 695–8.

Ein, M.E., Smith, N.J., Aruffo, J.F., Heererma, M.S., Bradshaw, M.W. and Williams, T.W. (1979). Susceptibility and synergy studies of methicillin-resistant *Staphylococcus epidermidis. Antimicrob. Agents Chemother.* **16,** 655–9.

Elwell, L.P., Roberts, M., Mayer, L.W. and Falkow, S. (1977). Plasmid mediated Beta-lactamase production in *Neisseria gonorrhoeae. Antimicrob. Agents Chemother.* **11,** 528–33.

Elwell, L.P., Saunders, J.R., Richmond, M.H. and Falkow, S. (1977). Relationships among some R-plasmids found in *Haemophilus influenzae. J. Bacteriol.* **131,** 356–62.

Eriquez, L.A. and D'Amato, R.F. (1979). Purification by affinity chromatography and properties of a *β*-lactamase isolated from *Neisseria gonorrhoeae. Antimicrob. Agents Chemother.* **15,** 229–34.

Escheverria, P., Ulyango, C.B., Ho, M.T., Verhaert, L., Komalarinia, S., Orskov, F. and Orskov, I. (1978). Antimicrobial resistance and enterotoxin production among isolates of *Escherichia coli* in the Far East. *Lancet* **ii,** 589–92.

Evans, R.T. and Taylor-Robinson, D. (1978). The incidence of tetracycline resistant strains of *Ureaplasma urealyticum. J. Antimicrob. Chemother.* **4,** 57–63.

Farrar, W.E. and Eidson, M. (1971). Antibiotic resistance in Shigella mediated by R-factors. *J. Infect. Dis.* **123,** 477–84.

Feldman, H.A. (1967). Sulfonamide resistant meningococci. *Amer. Rev. Medicine* **18,** 495–504.

Fu, K.P. and Neu, H.C. (1979). Inactivation of β-lactam antibiotics by *Legionella pneumophilia. Antimicrob. Agents Chemother.* **16,** 561–4.

Goldman, P.L. and Petersdorf, R.G. (1979). Significance of methicillin tolerance in experimental endocarditis. *Antimicrob. Agents Chemother.* **15,** 802–6.

Hall, W.H., Schierl, E.A. and Maccani, J.E. (1979). Comparative susceptibility of penicillinase-positive and negative *Neisseria gonorrhoeae* to 30 antibiotics. *Antimicrob. Agents Chemother.* **15,** 562–7.

Hammond, G.W., Lian, C.J., Wilt, J.C. and Ronald, A.R. (1978). Antimicrobial susceptibility of *Hemophilus ducreyi. Antimicrob. Agents Chemother.* **13,** 608–12.

Hedges, R.W., Vialard, J.L., Pearson, N.J. and O'Grady, F. (1977). R-plasmids from Asian strains of *Vibrio cholerae. Antimicrob. Agents Chemother.* **11,** 585–8.

Horodniceanu, T., Bougueleret, L., El-Solh, N., Bieth, G. and Delbos, F. (1979). High-level, plasmid-borne resistance to gentamicin in *Streptococcus faecalis* subsp. zymogenes. *Antimicrob. Agent Chemother.* **16,** 686–9.

Hyder, S.L. and Streitfeld, M.M. (1978). Transfer of erythromycin resistance from clinically isolated lysogenic strains of *Streptococcus pyogenes* via their endogenous phage. *J. Infect. Dis.* **138,** 281–6.

Imsande, J. (1978). Genetic regulation of penicillinase synthesis in gram-positive bacteria. *Microbiol. Reviews* **42,** 67–83.

Jacobs, M.R., Koornhof, H.J., Robins-Browne, R.M., Stevenson, C.M., Vermoak, Z.A., Freiman, I., Miller, G.B., Witcomb, M., Isaacson, M., Ward, J.I. and Austrian, R. (1978). Emergence of multiply resistant pneumococci. *New England J. Med.* **299,** 735–40.

Jacoby, G.A. (1980). Plasmid determined resistance to carbenicillin and gentamicin. In *Plasmids and Transposons,* ed. C. Stuttard and K.R. Rozee, pp. 83–93. Academic Press, New York.

Jorgensen, J.H., Crawford, S.A. and Alexander, G.A. (1980). Comparison of moxalactam (LY127935) and cefotaxime against anaerobic bacteria. *Antimicrob. Agents Chemother.* **17,** 901–4.

Karmali, M.A. and Fleming, P.C. (1978). Campylobacter enteritis. *Canadian Medical Assoc. J.* **120,** 1525–32.

Kayser, F.H. (1975). Methicillin-resistant Staphylococci. *Lancet* **ii,** 650–2.

Kim, K.S., Yoshimori, R.N., Imagawa, D.T. and Anthony, B.F. (1979). Importance of medium in demonstrating penicillin tolerance by group B streptococci. *Antimicrob. Agents Chemother.* **16,** 214–6.

Korzeniowski, O.M., Wennersten, C., Moellering, R.C. and Sande, M.A. (1978). Penicillin-netilmicin synergism against *Streptococcus faecalis. Antimicrob. Agents Chemother.* **13,** 430–4.

Lacey, R.W. (1975). Antibiotic resistance plasmids of *Staphylococcus aureus* and their clinical importance. *Bacteriol. Rev.* **39,** 1–32.

Laufs, R., Kaulfers, P.M., Jahn, G. and Teschner, U. (1979). Molecular characterization of a small *Haemophilus influenzae* plasmid specifying β-lactamase and its relationship to R-factors from *Neisseria gonorrhoeae. J. Gen. Microbiol.* **111,** 223–31.

Leung, T. and Williams, J.D. (1978). β-lactamases of subspecies of *Bacteroides fragilis. J. Antimicrob. Chemother.* **4,** (suppl. B), 47–54.

Markowitz, S.M. (1980). Isolation of an ampicillin-resistant, non-β-lactamase-producing strain of *Haemophilus influenzae. Antimicrob. Agents Chemother.* **17,** 80–3.

McCarthy, L.R., Mickelsen, P.A. and Smith, E.G. (1979). Antibiotic susceptibility of *Haemophilus vaginalis (Corynebacterium vaginale)* to 21 antibiotics. *Antimicrob. Agents Chemother.* **16,** 186–9.

McClatchy, J.K. (1980). Antituberculous drugs: mechanisms of action, resistance, susceptibility testing, and assays of activity in biological fluids. In *Antibiotics in Laboratory Medicine,* ed. V. Lorian, pp. 135–69. Williams and Williams, Baltimore.

Meyer, P.W. and Lerman, S.L. (1980). Rise and fall of Shigella antibiotic resistance. *Antimicrob. Agents Chemother.* **17,** 101–2.

Mhalu, F.S., Mmari, P.W. and Ijumba, J. (1979). Rapid emergence of El Tor Vibrio cholera resistant to antimicrobial agents during first six months of fourth cholera epidemic in Tanzania. *Lancet* **i,** 345–7.

Moodie, H.L. and Woods, D.R. (1973). Anaerobic R-factor transfer in *Escherichia coli. J. Gen. Microbiol.* **76,** 437–40.

Murray, B. and Moellering, R.C. (1978). Patterns and mechanisms of antibiotic resistance. *Medical Clinics N. America* **62,** 899–923.

Norlander, L., Davies, J. and Normark, S. (1979). Genetic exchange mechanisms in *Neisseria gonorrhoeae. J. Bacteriol.* **138,** 756–61.

Percheson, P.B. and Bryan, L.E. (1979). Penicillin binding proteins of penicillin-susceptible and resistant *Streptococcus pneumoniae.* In *Current Chemotherapy and Infectious Disease,* ed. J.D. Nelson and C. Grassi, pp. 703–4. American Society for Microbiology, Washington, DC.

Prince, A. and Neu, H.C. (1976). Beta-lactamase activity in *Shigella sonnei. Antimicrob. Agents Chemother.* **9,** 776–9.

Raevuori, M., Harvey, S.M., Pickett, M.J. and Martin, W.J. (1978). *Yersinia enterocolitica: in vitro* antimicrobial susceptibility. *Antimicrob. Agents Chemother.* **13,** 888–90.

Raynor, R.H., Scott, D.F. and Best, G.K. (1979). Oxacillin-induced lysis of *Staphylococcus aureus. Antimicrob. Agents Chemother.* **16,** 134–40.

Rennie, R.P. and Duncan, I.B.R. (1977). Emergence of gentamicin-resistant klebsiella in a general hospital. *Antimicrob. Agents Chemother.* **11,** 179–84.

Richmond, A.S., Simberkoff, M.S., Schaefler, S. and Rahal, J.J. (1977). Resistance of *Staphylococcus aureus* to semisynthetic penicillins and cephalothin. *J. Infect. Dis.* **135,** 108–12.

Roberts, M., Elwell, L.P. and Falkow, S. (1977). Molecular characterization of two Beta-lactamase-specifying plasmids isolated from *Neisseria gonorrhoeae. J. Bacteriol.* **131,** 557–63.

Ross, S., Controni, G. and Khan, W. (1972). Resistance of Shigellae to ampicillin and other antibiotics. *J. Amer. Med. Assoc.* **221,** 45–7.

Sabath, L.D., Laverdiere, M., Wheeler, N., Blazevic, D. and Wilkinson, B.J. (1977). A new type of penicillin resistance of *Staphylococcus aureus.* Lancet **i,** 443–8.

Sack, D.A., Kaminsky, D.C., Sack, B., Itotia, J.N., Arthur, R.R., Kapikian, A.Z., Orskov, F. and Orskov, I. (1978). Prophylactic doxycycline for travelers diarrhea. *New England J. Med.* **298,** 758–63.

Sadowski, P.L., Peterson, B.C., Gerding, D.N. and Cleary, P.P. (1979). Physical characterization of ten R-plasmids obtained from an outbreak of nosocomial

Klebsiella pneumoniae infections. *Antimicrob. Agents Chemother.* **15,** 616–24.

Scheifele, D.W. (1979). Ampicillin-resistant *Hemophilus influenzae* in Canada: nation-wide survey of hospital laboratories. *Canadian Med. Assoc. J.* **121,** 198–202.

Scudamore, R.A., Beveridge, T.J. and Goldner, M. (1979). Penetrability of the outer membrane of *Neisseria gonorrhoeae* in relation to acquired resistance to penicillin and other antibiotics. *Antimicrob. Agents Chemother.* **15,** 820–7.

Shoemaker, N.B., Smith, M.D. and Guild, W.R. (1979). Organization and transfer of heterologous chloramphenicol and tetracycline resistance genes in pneumococcus. *J. Bacteriol.* **139,** 432–41.

Siegel, M.S., Thornsberry, C., Biddle, J.W., O'Mara, P.R., Perine, P.L. and Wiesner, P.J. (1978). Penicillinase-producing *Neisseria gonorrhoeae:* results of surveillance in the United States. *J. Infect. Dis.* **137,** 170–5.

Sparling, P.F., Guymon, L. and Biswas, G. (1976). Antibiotic resistance in the gonococcus. In *Microbiology 1976,* ed. D. Schlessinger, pp. 494–500. American Society for Microbiology, Washington.

Speller, D.C. (1980). Hospital infection by multi-resistant gram-negative bacilli. *J. Antimicrob. Chemother.* **6,** 168–70.

Stuy, J.H. (1979). Plasmid transfer in *Haemophilus influenzae. J. Bacteriol.* **139,** 520–9.

Stuy, J.H. (1980). Chromosomally integrated conjugative plasmids are common in antibiotic-resistant *Haemophilus influenzae. J. Bacteriol.* **142,** 925–30.

Tally, F.P., Snydman, D.R., Gorbach, S.L. and Malany, N.H. (1979). Plasmid-mediated, transferable resistance to clindamycin and erythromycin in *Bacteroides fragilis. J. Infect. Dis.* **139,** 83–8.

Tally, F.P. (1978). Factors affecting antimicrobial agents in an anaerobic abscess. *J. Antimicrob. Chemother.* **4,** 299–302.

Tanaka, T., Ikemura, K., Tsunada, M., Sasagawa, I. and Mitsuhashi, S. (1976). Drug resistance and distribution of R-factors in Salmonella strains. *Antimicrob. Agents Chemother.* **9,** 61–4.

Thornsberry, C., Baker, C.N. and Jones, R.N. (1974). *In vitro* antimicrobial of piperacillin and seven other *β*-lactam antibiotics against *Neisseria gonorrhoeae* and *Haemophilus influenzae* including *β*-lactamase producing strains. *J. Antimicrob. Chemother.* **5,** 137–42.

Thornsberry, C., Baker, C.N. and Kirven, L.A. (1978). *In vitro* activity of antimicrobial agents on Legionnaires disease bacterium. *Antimicrob. Agent Chemother.* **13,** 78–80.

Threlfall, E.J., Ward, L.R. and Rowe, B. (1978). Epidemic spread of a chloramphenicol-resistant strain of *Salmonella typhimurium* phage type 204 in bovine animals in Britain. *The Veterinary Record* **103,** 438–40.

Timoney, J.F. (1978). The epidemiology and genetics of antibiotic resistance of *Salmonella typhimurium* isolated from diseased animals in New York. *J. Infect. Dis.* **137,** 67–73.

Tomich, P.K., An, F.Y., Damle, S.P. and Clewell, D.B. (1979). Plasmid-related transmissibility and multiple drug resistance in *Streptococcus faecalis* subsp. *zymogenes* strain DS16. *Antimicrob. Agents Chemother.* **15,** 828–30.

Tompkins, L.S., Plorde, J.J. and Falkow, S. (1980). Molecular analysis of R-factors from multiresistant nosocomial isolates. *J. Infect. Dis.* **140,** 625–36.

Towner, K.J., Pearson, N.J. and O'Grady, F. (1979). Resistant *Vibrio cholerae* El Tor in Tanzania. *Lancet* **ii,** 147–8.

Walder, M. (1979). Susceptibility of *Campylobacter fetus* subsp. *jejuni* to twenty antimicrobial agents. *Antimicrob. Agents Chemother.* **16,** 37–9.

Weidner, C.E., Dunkel, T.B., Pettyjohn, F.S., Smith, C.D. and Leibovitz, A. (1971). Effectiveness of rifampin in eradicating the meningococcal carrier state in a relatively closed population: Emergence of resistant strains. *J. Infect. Dis.* **124,** 172–8.

Wilkinson, A.E. (1977). The sensitivity of gonococci to penicillin. *J. Antimicrob. Chemother.* **3,** 197–8.

Yourassowsky, E., Van, DerLinden, M.P. and Lismont, M.J. (1979). Growth curves, microscopic morphology, and subcultures of Beta-lactamase-positive and negative *Haemophilus influenzae* under the influence of ampicillin and cefamandole. *Antimicrob. Agents Chemother.* **15,** 325–31.

7

The control of antibiotic resistance

Development and dissemination of antibiotic resistance

Antibiotic use and the development of resistance

The use of antimicrobial agents and the emergence of antibiotic resistance are intimately related. It has been shown on several occasions that use and misuse of various antibiotics has been associated with the emergence of resistant bacteria. Resistant organisms have been R-factor containing derivatives or mutations of previously susceptible bacteria or naturally occurring intrinsically resistant bacteria. Although development of resistance has been strongly influenced by antibiotic usage, some R-factor specified resistance has been detected in bacteria rarely or never treated with that agent.

Emergence of resistance related to use of antibiotics can be considered for the individual course of therapy, the community and for an institution.

It is of interest that development of resistance during a course of systemic antimicrobial therapy for an infection is only uncommonly the cause of failure of treatment. However, some exceptions to this generalization do occur. Treatment of *Mycobacterium tuberculosis,* for example with streptomycin, has been associated with failure due to a streptomycin-resistant organism. Therapy of *Cryptococcus neoformans* meningitis with 5-fluorocytosine (5-FC) may often fail due to 5-FC resistant mutants. Treatment of *E. coli* urinary tract infections with nalidixic acid not uncommonly fails due to emergence of resistance. However, situations such as these are only occasionally documented.

Those cases where a specific therapeutic failure has been due to emergence of resistance in the organism being treated most often involve the endocardium, the meninges or the kidneys. These are infections in which host defenses are considered to be of less value in clearing the infecting microbe than for many other sites of infection. It is possible that host mechanisms may have to be of reduced efficacy to allow growth of resistant organisms from initially very low numbers. Resistance mechanisms particularly associated with mutation may sometimes cause a reduction of virulence impairing the capability of resistant mutants to persist in tissues. Bacteria which acquire R-factors have not been shown to be associated with a significant decrease in virulence in most cases. Thus R-factor transfer to the primary pathogen may not occur frequently enough *in vivo* during an infection to allow persistence of R^+ derivatives in the face of normal host defense mechanisms.

A problem of greater importance has been the clear association of antibiotic use and the emergence of resistant bacteria among previously susceptible pathogenic bacteria from the general community. One of the most striking examples has been the development of penicillin and other antibiotic resistance among *Staphylococcus aureus* (see also Chapter 6). Although this evolution of resistance was initially more prominent in hospitals, it has subsequently involved the community. The capability of *S. aureus* to respond to the selective pressure of antibiotic use continues with the recent development of gentamicin resistance in many parts of the world. Ampicillin resistance of *Hemophilus influenzae* and penicillin resistance of *Neisseria gonorrhoeae* are other important examples. The use of sulfonamides for treatment of *Shigella* infections in Japan was associated with a marked increase in frequency of R-factors specifying multiple resistance. Chloramphenicol resistance of *Salmonella typhi* in Mexico and Southeast Asia is yet another important situation illustrating the relationship of drug use and community resistance. The patterns of resistance which have emerged among important community pathogens vary with geographic locations and with time.

The extensive use of tetracycline in many areas of the world has resulted in much resistance to that antibiotic. Significant resistance among major pathogens includes *S. aureus, Streptococcus pyogenes, S. pneumoniae, Bacteroides fragilis, N. gonorrhoeae* and on R-factors in many Enterobacteriaceae.

The development of antibiotic resistance within hospitals and sections of hospitals has been repeatedly observed. The major problem has been the selection of many bacteria of low pathogenicity but with intrinsic and/or acquired antibiotic resistance. Infections with these organisms can occur in hospitals because of patients with a variety of impaired host defenses.

Bacteria which have posed resistance problems in hospitals include *S. aureus, S. epidermidis, Proteus* species, *Pseudomonas aeruginosa,* other *Pseudomonas* species, *Providencia, Citrobacter, Enterobacter* spp., *Klebsiella* spp., *E. coli* and numerous other gram-negative bacilli.

A resistant strain (or strains) of a single species has frequently been shown to cause outbreaks of infections within a hospital or hospital ward. Such nosocomial organisms have most often been *Serratia, Providencia, Klebsiella, Pseudomonas, E. coli* and *Proteus*. They most frequently cause urinary tract and wound infections. The reservoir of resistance varies among the types of bacteria. It is often urinary or wound infections for *Serratia, Providencia, Pseudomonas* and *Proteus* but is probably the gut for Klebsiella and *Citrobacter* strains. The major factor in most nosocomial outbreaks is cross-infection of predisposed patients with a resistant strain or strains. However, the source of a resistant strain may vary being endogenous to the environment or being the result of molecular evolution of plasmids.

The development of resistance in a particular strain or strains involved in nosocomial infection may involve plasmid transfer and *in vivo* transposition of resistance. An outbreak of nosocomial infections has been described involving four hospitals and associated with a strain of *Serratia marcescens*. Analysis of the plasmid DNA of strains isolated during the outbreak showed that transposition of resistance for kanamycin, gentamicin, ampicillin and carbenicillin had occurred from a small nonconjugative plasmid to a large conjugative plasmid. This conjugative plasmid was partly responsible for dissemination of resistance to other bacteria involved in the outbreak.

The use of topical aminoglycosides, particularly neomycin and gentamicin, has been an important factor in the development of outbreaks of antibiotic resistance. R-factors of *P. aeruginosa* specifying gentamicin resistance emerged as a result of use of topical gentamicin in burn units. Restriction of gentamicin use has been shown to be associated with a marked reduction of gentamicin resistance in a hospital for burned patients.

Resistance to aminoglycosides among hospital strains of *E. coli, Klebsiella, Enterobacter* and *Serratia* is usually due to enzymic inactivation of the drugs. However most hospital examples of resistant *P. aeruginosa* infection except those associated with topical gentamicin use in burn units are due to strains which do not have inactivating enzymes. This type of resistance is less common in the Enterobacteriaceae but does occur. The control of these two forms of resistance is different. The enzymatic resistance is best controlled by barrier type precautions whereas the non-enzymatic form responds best to use of effective dosage of aminoglycosides and the limitation of such use to only essential cases.

The use of antibiotic prophylaxis can lead to infections with resistant organisms if the prophylaxis is continued for several days. However, the use of antibiotic prophylaxis for surgical procedures in which the drug is given to achieve adequate tissue levels only during surgery, has not been associated with significant increases in antibiotic resistance.

Persistence of R-factors in the human intestine can be enhanced by use of antibiotics. However, once established an *E. coli* strain with an R-factor has survived in the gut for several weeks in the absence of antibiotic use. The spread of R-factor specified tetracycline resistance in an *E. coli* strain to the spouse of a patient taking prolonged tetracycline has been demonstrated.

Source and dissemination of antibiotic resistance

It is difficult to present a total analysis of the sources of resistance because of the variation in resistance mechanisms. Some of the most important of these will be considered.

In hospitals antibiotic resistance is frequently the result of the selection of strains which possess considerable intrinsic resistance to antibiotics and may have superimposed acquired resistance which is most often due to plasmids. These strains are normally bacteria which are or can become part of the patient's bacterial flora. Patients who enter hospitals and remain for days to weeks undergo progressive colonization of various body sites but especially the gut, upper respiratory tract and moist skin areas with various gram-negative bacilli. These organisms represent many noted earlier in this chapter and are recognized as major causes of nosocomial infections. *S. epidermidis* is a gram-positive organism which is an increasing resistance problem particularly causing infections involving foreign bodies. Use of antibiotics can then select for these endogenously resistant strains or for endogenous strains which have acquired additional resistance most often through plasmid acquisition and in some cases transposition of resistance genes.

The original source of antibiotic resistance genes carried by transposons and plasmids is unknown. Bacteria that produce antibiotics frequently have resistance mechanisms to protect themselves. The production of inactivating enzymes for some aminoglycosides is shared by some producing strains and some strains with acquired resistance. For example the aminoglycoside phosphotransferase (APH(3′)) from *Bacillus circulans* has been introduced to *E. coli* and causes a typical resistance phenotype. *B. circulans* produces butirosin. However, antibodies do not cross-react with (APH(3′)) enzymes of resistant and producing strains.

Whatever the original source of resistance genes, they have been detected in bacteria isolated prior to the clinical use of antibiotics. Presumably bacteria have faced antibiotics in nature from naturally occurring antibiotic-producing microorganisms for eons of time.

The late emergence of ampicillin resistance in *Hemophilus influenzae* compared to enteric bacteria has prompted much speculation on the origin of this resistance. In general, two types of plasmids have been detected. One group has molecular weights of $30-38 \times 10^6$, is conjugative, shares substantial base sequence homology and has guanosine + cytosine (G + C) base composition of about 39%. The G + C composition is very similar to that of the *H. influenzae* chromosome. These plasmids carry the entire TnA base sequence.

The second type is $2.5-5 \times 10^6$ molecular weight, non-conjugative, has highly homologous (85–90%) base sequence and contains part of the TnA base sequences. These have been found in *N. gonorrhoeae* as well.

The evidence from *H. influenzae* suggests that the conjugative plasmid originated from *H. influenzae.* Recently it has been shown that such plasmids frequently integrate into the *H. influenzae* chromosome and can be apparently excised to become conjugative. The results of investigations to date are consistent with the view that the TnA gene for β-lactamase was transposed to a core cryptic conjugative plasmid in *H. influenzae.* It is very unlikely that a broad host range plasmid (promiscuous plasmid) carried the β-lactamase gene to *H. influenzae* from enteric bacteria and remained as a resident plasmid. The small plasmids are of unknown origin but their wide distribution suggests they may represent a gene pool for β-lactamase.

The origin of antibiotic resistance genes and their route between bacterial species remains obscure for most resistance development. The continued study of the molecular characteristics of plasmids involved in such resistance may ultimately solve this very difficult question. Such studies have lead to a better appreciation of the potential of plasmids and transposons for dissemination and development of resistance.

It is clear that extrachromosomal elements represent a potent force to maintain antibiotic resistance profiles of bacteria in a plastic state. Plasmids exist which have the capability to conjugate and transfer resistance between only closely related bacteria (e.g. P-2 plasmids of *Pseudomonas*) or between widely different bacteria (e.g. P plasmids). The capability to transfer resistance throughout most clinically important gram-negative bacteria exists. As well conjugative plasmids of *Streptococcus faecalis* have been described.

Even in bacteria without recognized conjugation, plasmid transfer has still been successful particularly by transduction but also by transformation. Witness the prevalence of plasmid-mediated resistance among staphylococci.

While plasmids are very important for transfer of resistance, it must also be remembered that many are unstable to varying degrees in their bacterial hosts. Thus resistance can be lost from bacteria. In some instances, reduced use of antibiotics has been associated with a reduction in the prevalence of plasmid-specified resistance.

Another very potent component of the evolution of antibiotic resistance is the large array of transposable DNA resistance phenotypes (see Chapter 4). As noted earlier in this chapter transposition *in vivo* has been shown to occur and result in a change in the transferable antibiotic resistance pattern of a strain of *S. marcescens. In vivo* transfer of an antibiotic resistance plasmid has also been documented in humans and in animals. These mechanisms for antibiotic resistance development and dissemination mean that the study of the epidemiology of resistant strains particularly in hospitals must increasingly take into account plasmid characteristics as well as those of strains.

The study of outbreaks of nosocomial infection should involve methods to identify strains of the species involved. These may be serotyping, phage typing, bacteriocin typing, biotyping and occasionally other methods. For some bacteria such methods are readily available but are less available for some of the uncommon gram-negative bacilli such as *Citrobacter.* If possible it is often of use to combine typing methods to increase the discrimination between strains. Studies of this type have shown dissemination of resistant strains through a ward or hospital on several occasions. Additional studies which should be increasingly used include: determination of resistance mechanisms, methods for transfer of an R-plasmid, incompatibility typing of a detected plasmid and agarose gel electrophoresis of cell lysates and of restriction endonuclease digests of extracted DNA. Methods for these and additional procedures are given in the Elwell and Falkow 1980 reference provided at the end of this chapter.

The use of the additional studies noted will frequently allow the determination of the extent of resistance transposition and plasmid transfer. Such studies have shown that dissemination of the gene for 2″ aminoglycoside nucleotidyltransferase (ANT(2″)) was by a M-group plasmid which transferred among various strains and species of Enterobacteriaceae rather than by transposition of the gene to different plasmids.

Spread of resistant strains does not necessarily involve plasmid-mediated resistance. Nosocomial outbreaks of *P. aeruginosa* in a Toronto hospital involved permeability type aminoglycoside resistance. This form of resistance is probably mutational or due to transduction of chromosomal genes.

In summary, development of antibiotic resistance may involve patient-to-patient spread of strains, transposition of DNA between plasmids and plasmid transfer among bacteria as well as mutational resistance. All have been shown to occur in outbreaks of nosocomial infection and probably occur in the dissemination of community resistance.

It is clear that the use of some antibiotics including tetracycline, chloramphenicol, sulfonamides, ampicillin and neomycin in animal feeds can select for multiple resistance. Indeed the administration of tetracycline to mice carrying resistant strains of *Bacteroides fragilis* can induce tetracycline resistance transfer. A major potential for selection and dissemination of antibiotic resistance exists in circumstances where animals are fed antibiotics.

Some investigations have shown that antibiotic resistance in animal *Salmonella* strains can be transmitted to humans. Calves fed tetracycline in rearing stations in Great Britain developed multiple resistance in *Salmonella typhimurium* phage type 29. Subsequently multiply resistant strains of the same type appeared in a human outbreak. Other studies have shown that resistance can develop in response to the therapeutic use of chloramphenicol in calves. In Great Britain a marked increase in multiply resistant *S. typhimurium* phage type 204 has spread throughout cattle. Some chloramphenicol resistant *S. typhimurium* of phage type 204 have caused human disease.

In situations examined to date antibiotic resistance in humans from probable animal sources has involved spread of strains from animals to humans rather than plasmid spread from animal to human *Salmonella* strains. In most cases studied the bacteria were Salmonellae. Once established in humans these strains characteristically spread among human populations. The spread of plasmids from animal to human strains could also occur but there is little evidence to support it. Thus, the IncH group plasmids responsible for chloramphenicol resistance in *S. typhi* are more likely to have resulted from the widespread use of this drug for humans in some parts of the world. It is unlikely that these plasmids originated in animal strains of *Salmonella*. The major risk to date has been the development of multiple antibiotic resistance in those animal *Salmonella* strains which can readily infect humans.

Prevention and elimination of antibiotic resistance

Reduction of the selective pressure of antibiotic use

Antimicrobial agents are a very widely used group of drugs both in human and animal medicine and in rearing animals for commercial food purposes.

Numerous reports have documented the overuse and misuse of antibiotics in adult and pediatric medicine. Development of antibiotic resistance is a clear risk of this situation but excess drug reactions, toxicity and cost are other very serious results of antibiotic overuse.

Reduced and enlightened use of antibiotics is probably most influenced by improved medical education of students, residents and graduate physicians in the basic and clinical aspects of infectious disease and antibiotic use. The tremendous input of pharmaceutical firms into drug advertisement and sales is not a proper source of education for the most part. In fairness some firms have assisted sponsorship of educational exercises unrelated to their products. A program to improve the public's understanding of the ill results of antibiotic misuse could materially assist in reducing pressure on physicians to prescribe drugs.

The principles guiding use of antibiotics should not be really any different from those for other classes of potent and often expensive drugs. A correct diagnosis, isolation of causative bacteria and accurate antibiotic susceptibility testing are important to guide initial drug selection or retrospectively to confirm correct selection of an agent. A definite indication for antibiotic use should exist before antibiotics are prescribed. The prophylactic administration of agents to individuals with trivial viral respiratory infections and many other frequent excuses for antibiotic use are not indications.

Some specific forms of drug use should be curtailed. Antibiotic prophylaxis is effective in certain circumstances. Its use in association with surgical procedures should be designed so that the agent is administrated only long enough to provide adequate tissue levels at the operative site during the surgical procedure. This requires only one dose in most cases and rarely more than three.

Topical antibiotic agents should be avoided where possible. If used they should not include agents which are also used systemically. Some drugs such as bacitracin, polymyxins, vancomycin, nitrofurans and amphotericin B are rarely associated with resistance. Agents like aminoglycosides and β-lactams are frequently associated with resistance.

Drug resistances not currently found on R-factors or producing insignificant resistance can be used if indicated as their use reduces the selective pressure for multiple resistance. Drugs like nitrofurans may be indicated for situations where prolonged treatment of lower urinary tract infections is required. If prolonged therapy is necessary the use of two or sometimes more agents can prevent or reduce resistance. Examples are multiple drug therapy of tuberculosis and the use of co-trimoxazole for recurrent lower urinary tract infections.

Restriction of the use of those antibiotics used for human infections from animal feeds or extensive animal therapeutic programs should be strongly considered. There is an increasing tendency to use agents relatively specific for gram-positive bacteria. While this may help decrease resistance among gram-negatives, it raises the spectre of more resistance in gram-positives.

Restriction of the use of specific newly introduced antibiotics within a hospital is probably best approached on a cost basis. There have been several examples of attempts to restrict drug availability to reduce antibiotic resistance. Amikacin is an agent subject to these restrictions in many institutions based on its resistance to many of the aminoglycoside-inactivating enzymes. The argument for restriction has been that it will select for those enzymes which inactivate it. This argument can be reversed to state it will select for very few of the enzymes and thus may be a preferential agent. Any of the modern aminoglycosides select for permeability type resistance. The better way to curtail aminoglycoside resistance is to reduce the use of all of these agents.

New antibiotics are generally more costly than those in prior use. This is a strong argument for restriction of their use to specific treatment circumstances. If cost is equal, a new agent may be preferable. Antibiotics resistant to β-lactamases will be unlikely to select for β-lactamase resistance. Rather they may select for permeability and target type resistance. In general for most antibiotics (but important exceptions exist e.g. streptomycin), these forms of resistance are of a lower level, are mutational and may sometimes be associated with reduced bacterial virulence. Thus β-lactamase resistant β-lactams have potential advantages over their susceptible counterparts.

The use of a narrow spectrum drug in preference to a broad spectrum agent if both are effective has been stated as a method to reduce resistance. While this seems reasonable, the evidence to support it as an important way to reduce resistance is generally lacking.

Another possible way to control antibiotic resistance through drug use patterns is to rotate the use of antibiotics. Rotation of the use of drugs particularly with different targets is a potential way to reduce resistance. This procedure has not been extensively tried. However, with the increased number of new agents, the necessary armamentarium of drugs is available to try it within hospitals and perhaps communities.

Recognition of different antibiotic resistance patterns and control of their spread

Detection of new resistance patterns in the community or the hospital is facilitated by active surveillance programs. In hospitals these usually involve the hospital infection control officer and the microbiology laboratory. Some hospitals have sophisticated computerized programs to recognize different antibiotic resistance patterns. Whatever the type of program instituted, it requires accurate bacteriological identification and antibiotic susceptibility testing. Heavy emphasis should be placed on the detection of any clusters of infections and unusual resistance patterns.

In communities, resistance is often recognized initially by the bacteriological assessment of treatment failures. Large-scale surveillance by public health laboratories is of particular value to assess the significance of a resistance problem in the community. Thus, for example, the Public Health Laboratory of Alberta has examined *S. pneumoniae* for various forms of antibiotic resistance. They determined that 2.2% of 8654 isolates were resistant to eight-fold or greater levels of penicillin than those required to inhibit control strains.

Assessment of ampicillin resistance in *H. influenzae,* penicillin resistance in *N. gonorrhoeae,* chloramphenicol and ampicillin resistance of *S. typhi,* multiple resistance of *Shigella,* resistance patterns of *M. tuberculosis* are other examples of surveillance information of major value to treat important infections.

Methods to reduce spread of resistance are difficult to institute effectively in the community. The spread of β-lactamase resistance in *N. gonorrhoeae* illustrates the problem. This form of resistance now seems world-wide. Its control depends to a large extent on the control of gonorrhoea. That has had poor success in most countries of the world. All strains of *N. gonorrhoeae* should ideally be tested for β-lactamase and the eradication of producing strains insured. At the very least, all treatment failure strains should be examined and eradicated. Public health officials should place major effort on the follow-up of patients with β-lactamase producing strains of *N. gonorrhoeae.*

In general, resistance control can be centered on an accurate and sensitive system to detect resistance. Thereafter control of these index strains can be relentlessly pursued. Alternatively the control of spread of a pathogenic bacterium through the community can be the focus of activity. This has been successful to varying degrees for tuberculosis, although many factors in the community contribute to its control. The control of gonorrhoea has, for the most part, been a failure. Control of spread of β-lactamase specified resistance has also been largely a failure.

Control of spread of resistance in hospitals has been reasonably successful in many institutions. This requires the capability to detect resistance and the determination of reservoirs of resistant strains, routes of dissemination between patients and antibiotic use patterns maintaining antibiotic resistant strains in the local environment.

Once the hurdle of recognizing that a resistance problem exists in the hospital is overcome, control is often surprisingly straightforward. There are too many circumstances to discuss each specifically but some can be mentioned. Time spent in observation of procedures carried out in the involved wards is usually most rewarding. For example breaks in technique involving closed urinary catheter drainage frequently explain an outbreak of *Providencia*-mediated urinary tract infections. Attention to the cleaning of possible reservoirs such as nebulization chambers reduced *Pseudomonas* infections.

If necessary better barrier techniques between patient and patient and between patient and staff may be needed. This is especially so with patients with compromised host defenses. Barrier techniques can vary from simple (e.g. greater use of gloves, bagging sputum and wound secretions, segregating patients with indwelling devices especially urinary catheters, etc.) to very sophisticated (e.g. laminar air flow rooms, total reverse isolation). Most problems can be resolved by simple methods. Sophisticated techniques are needed only for those with profound deficiencies of host defenses.

Antibiotic use patterns should be reviewed. Correlation of increased resistance and increased drug use can be found in many instances if carefully assessed. Some forms of resistance can only be controlled by altered drug use. An important example is permeability type aminoglycoside resistance which may develop individually in each patient involved.

The route of spread of resistance can sometimes involve several types of strains and different species. The methods for tracking plasmid spread noted earlier should be used under these circumstances.

Resistance and new antibiotics

New agents, antibiotic modification and non-antibiotic inhibitors of resistance mechanisms as an approach to overcome resistance

There are numerous ways to attempt to circumvent resistance by the use of new or modified antibiotics. The search for new effective naturally produced antibiotics has had increasingly diminished returns in recent years although occasional agents are still found in this manner. Important examples are thienamycin and cephamycins. Even these have been chemically modified for clinical use. Modification of agents previously felt to be too toxic or subject to reversal of action has yielded some new agents e.g.

deuterated fluoro-D-alanine. The major route for new antibiotics including those resistant to resistance mechanisms has been the modification of previous agents particularly β-lactams and aminoglycosides.

Modification of 6-aminopenicillanic acid, 7-aminocephalosporanic acid and the cephamycin nucleus has been a very productive route for new penicillins, cephalosporins and cephamycins. Many cephalosporin derivatives have been synthesized with significantly increased resistance to destruction by a variety of β-lactamases. These include cefamandole, cefuroxime cefotaxime, cefsulodin and cefoperazone.

Cephamycins are 7-α-methoxy cephalosporins. This substitution at the 7 position much reduces the affinity of these agents for a wide variety of β-lactamases. They have higher resistance to hydrolysis by β-lactamases than the group noted above. They do not inhibit β-lactamases because of their low enzyme affinity. Examples are cefoxitin and cefmetazole. Moxalactam contains an oxygen at the 1 position of the cephalosporin nucleus and has a 7-α-methoxy side chain. This agent is also very resistant to β-lactamases and is usually two to four or more times as active as cefoxitin on gram-negative bacteria. Presumably it penetrates the outer membrane of gram-negative bacteria more effectively than cefoxitin.

Some modified cephalosporins have reasonably good antipseudomonal activity including moxalactam, cefulodin, cefoperazone and cefotaxime. However this is not achieved primarily through β-lactamase resistance but rather by better cell entry.

Modification of the penicillin nucleus has been used to produce many different antibiotics. A group of these have increased resistance to staphylococcal β-lactamases and can act as inhibitors for certain enzymes (see Chapter 3). These include methicillin, isoxazoyl penicillins and nafcillin. Many other semisynthetic penicillins have reduced susceptibility to a variety of β-lactamases of gram-negative bacteria although they all are hydrolyzed. These include carbenicillin, ticarcillin, sulbenicillin, piperacillin, azlocillin, mezlocillin and other agents. Mecillinam is a 6-amidinopenicillanic acid derivative. It has a high affinity for PBP-2 of *E. coli* and penetrates gram-negative cell walls very effectively.

Other penicillin derivatives do not possess major resistance to β-lactamase or significantly enhance cell penetration.

Thienamycins and olivanic acids are naturally occurring β-lactams produced by streptomyces which are antibiotics and inhibit many different β-lactamases. Derivatives of these compounds are currently under investigation. A formamide derivative of thienamycin has a very broad spectrum of activity including *Pseudomonas* but does not attack penicillin-resistant pneumococci, some gram-positive bacilli and *Pasteurella multocida.* This thienamycin derivative is more active and more stable than the parent compound. Its intrinsic activity is very high for many different types of bacteria.

Several β-lactams which act as inhibitors of β-lactamases but are not antibiotics are at various stages of assessment. Clavulanic acid inhibits most β-lactamases of gram-positive and negative bacteria, including TEM and OXA plasmid-mediated enzymes and *Bacteroides fragilis, Klebsiella, Proteus* and *S. aureus* enzymes. It is less effective against many cephalosporinases. It is under assessment combined with the β-lactamase susceptible β-lactam, amoxycillin. It is distributed similarly to amoxycillin in tissues. This is important as it has no intrinsic antibiotic activity. Other β-lactamase inhibitors of this type include the semisynthetic compounds penicillanic acid sulfone (CP45899) and β-bromopenicillanic acid.

Aminoglycosides have also been extensively modified to yield agents active on bacteria possessing inactivating enzymes. In general it is probable that most of these agents act by having reduced affinity for the inactivating enzyme. Thus they are transported into cells and bind to ribosomes before significant inactivation can occur.

The 1-amino group of the deoxystreptamine ring of these agents can be acylated or alkylated in many cases without loss of activity but with reduced susceptibility to inactivating enzymes. Amikacin is substituted at the 1-amino group with L(−)-α-amino-α-hydroxybutyric acid. It is only effectively inactivated by AAC(6′) and ANT(4′) enzymes (see Chapter 3). Other 1-amino derivatives with reduced susceptibility to inactivating enzymes include netilmicin (1-*N*-ethyl sisomicin) and UK31213 (1-*N*-

dihydroxyisopropyl kanamycin B). Other effective modifications include the 5-epi-derivatives of sisomicin and gentamicin B, 3′,4′-dideoxykanamycin B, 6′-*N*-methyl-3′,4′,-dideoxykanamycin B and several others.

Sorbistin A, is a naturally produced aminoglycoside with a linear aglycone chains. It is not susceptible to known aminoglycoside inactivating enzymes.

Unfortunately the new aminoglycosides whether natural or semisynthetic do not have increased activity on strains with permeability type resistance. Most of these have been *P. aeruginosa* but some are Enterobacteriaceae.

Fluorinated analogs of chloramphenicol and thiamphenicol are active on bacteria producing chloramphenicol acetyltransferase. The drug 3-fluoro-3-deoxy-chloramphenicol (Sch24893) and similar derivatives of thiamphenicol (Sch25298) and fluorthiamphenicol (Sch25393) are active against a wide spectrum of bacteria but not including *Pseudomonas* and *Serratia.* The serious bone marrow suppressive toxicity of chloramphenicol needs to be assessed for these compounds.

Many other compounds are under assessment for their activity on resistant bacteria. These include rosoxacin (synthetic), bicyclomycin (natural), tetroxoprim (synthetic 2,4-diaminopyrimidine), fortimicin A (a natural aminoglycoside) and phosphonic acid derivatives.

A potential problem to the success of agents resistant to inactivating resistance mechanisms is the development of alternative modes of resistance. This situation has already been seen in *P. aeruginosa* where permeability resistance to aminoglycosides is widespread. It is possible that permeability and insusceptible target (e.g. penicillin-resistant pneumococci) types of resistance may be selected by wide use of the preceding antibiotics.

A series of additional approaches to reduce the prevalence of particularly R-factor specified resistance exist. Many of these are impractical because they would have to be effective on 100% of the bacterial population and human populations in many cases. These approaches could be of use in selected individual cases. They include drugs which act through R-factor specified pili such as macarbomycin, curing agents, agents inhibiting conjugation and genetic transfer, and drugs acting on R-factor specific replication events. Colonization of patients or specific tissues of patients in relatively closed environment such as an intensive care unit with sensitive bacteria is another possible method of limited potential application.

Another route of approach would be to devise drugs which when acted on by R-factor enzymes become toxic to the bacterial host. This would be analogous to the reduction of metronidazole by electron transport components of anaerobic bacteria to form intermediates toxic to the bacterial cell.

Antibiotics which are totally synthetic generally do not have the level of antibiotic resistance as natural products. In particular R-factor mediated resistance to agents like nalidixic acid and nitrofurantoin is not found or is ineffectual. Synthetic agents may present unique problems to bacteria for development of resistance mechanisms in that drug and bacteria may never have met before.

Selected references

Anderson, E.S. (1969). Ecology and epidemiology of transferable drug resistance. *Ciba Foundation Symposium. Bacterial Episomes and Plasmids*, pp. 102–15. Churchill Livingstone, London.

Anderson, E.S. (1977). The geographical predominance of resistance transfer systems of various compatibility groups in Salmonellae. In *R-factors: Their Properties and Possible Control,* ed. J. Drews and G. Högenauer, pp. 25–38. Springer-Verlag, Vienna.

Anderson, J.D., Gillespie, W.A. and Richmond, M.H. (1973). Chemotherapy and antibiotic-resistant transfer between Enterobacteria in the human gastro-intestinal tract. *J. Med. Microbiol.* **6,** 461–73.

Bryan, L.E. and Van Den Elzen, H.M. (1977). Spectrum of antibiotic resistance in clinical isolates of *Pseudomonas aeruginosa.* In *Microbiology 1977,* ed. D. Schlessinger, pp. 164–8. American Society for Microbioliology, Washington.

Bryan, L.E., Haraphongse, R. and Van Den Elzen. (1976). Gentamicin resistance in clinical-isolates of *Pseudomonas aeruginosa* associated with diminished gentamicin accumulation and no detectable enzymatic modification. *J. Antibiotics* **29,** 743–853.

Buckwold, F.J. and Ronald, A.R. (1979). Antimicrobial misuse – effects and suggestions for control. *J. Antimicrob. Chemother.* **5,** 129–36.

Davies, J. and Smith, D.I. (1978). Plasmid-determined resistance to antimicrobial agents. *Ann. Review Biochem.* **32,** 469–518.

Dixon, J.M.S., Lipinski, A.E. and Graham, M.E.P. (1977). Detection and prevalence of pneumococci with increased resistance to penicillin. *Canadian Med. Assoc. J.* **117,** 1159–61.

Elwell, L.P. and Falkow, S. (1980). The characterization of plasmids that carry antibiotic resistance genes. In *antibiotics in Laboratory Medicine,* ed. V. Lorian, pp. 433–53. Williams and Wilkins, Baltimore.

Elwell, L.P., Saunders, J.R., Richmond, M.H. and Falkow, S. (1977). Relationship among some R-plasmids found in *Haemophilus influenzae. J. Bacteriol.* **131,** 356–62.

Falkow, S. (1975). *Infectious Multiple Drug Resistance,* chapters 9 and 10. Pion Ltd, London.

Finland, M. (1972). Changing patterns of susceptibility of common bacterial pathogens to antimicrobial agents. *Annals Internal Medicine* **76,** 1009–36.

Grassi, G.G. (1977). Clinical aspects of the relationship between antibiotic usage and resistance. *J. Antimicrob. Chemother.* **3,** (suppl. C), 77–84.

Hirschmann, J.V. and Inui, T.S. (1980). Antimicrobial prophylaxis: a critique of recent trials. *Reviews Infectious Dis.* **2,** 1–23.

Jacob. A.E. and Hobbs, S.J. (1974). Conjugal transfer of plasmid-borne multiple antibiotic resistance in *Streptococcus faecalis* var *zymogenes. J. Bacteriol.* **117,** 360–72.

Miller, G.H., Arcieri, G., Weinstein, M.J. and Waitz, J.A. (1976). Biological activity of netilmicin, a broad-spectrum semisynthetic aminoglycoside antibiotic. *Antimicrob. Agents Chemother.* **10,** 827–36.

Mouton, R.P., Glerum, J.H. and van Loenen, A.C. (1976). Relationship between antibiotic consumption and frequency of antibiotic resistance of four pathogens (a seven-year survey). *J. Antimicrob. Chemother.* **2,** 9–19.

Naito, T., Nakagawa, S., Fujisawa, K. and Kawaguchi, H. (1975). Structure-activity relationships in amikacin analogs. In *Microbiol Drug Resistance,* ed. S. Mitsuhashi, pp. 425–39. University of Tokyo Press, Tokyo.

Nelson, J.D. and Grassi, C. (eds.) (1970). *Current Chemotherapy and Infectious Disease,* pp. 62–480. American Society for Microbiology, Washington.

Novick, R.P. and Morse, S.I. (1967). *In vivo* transmission of drug resistance factors between strains of *Staphylococcus aureus. J. Exp. Med.* **125,** 45–59.

O'Brien, T.F., Ross, D.G., Guzman, M.A., Medeiros, A.A., Hedges, R.W. and Botstein, D. (1980). Dissemination of an antibiotic resistance plasmid in hospital patient flora. *Antimicrob. Agents Chemother.* **17,** 537–43.

O'Callaghan, C.H. (1979). Description and classification of the newer cephalosporins and their relationships with the established compounds. *J. Antimicrob. Chemother.* **5,** 635–71.

Olsen, R.H. and Shipley, P. (1973). Host range and properties of the *Pseudomonas aeruginosa* R-factor 1822. *J. Bacteriol.* **113,** 772–80.

Pohl, P. (1977). Relationship between animal feeding in animals and emergence of bacterial resistance in man. *J. Antimicrob. Chemother.* **3** (suppl. C), 67–72.

Petrocheilou, V., Richmond, M.H. and Bennett, P.M. (1979). Persistence of plasmid-carrying tetracycline-resistant *Escherichia coli* in a married couple, one of whom was receiving antibiotics. *Antimicrob. Agents Chemother.* **16,** 225–30.

Privitera, G., Sebald, M. and Fayolle, F. (1977). Common regulatory mechanism of expression and conjugative ability of a tetracycline resistant plasmid in *Bacteroides fragilis. Nature* (Lond.) **278,** 657–8.

Richmond, M.H. (1977). Measures against the spread of R-factors (round-table discussion). In *R-factors: Their Properties and Possible Control,* ed. J. Drews and G. Högenauer, pp. 335–42, Springer-Verlag, Vienna.

Rolinson, G.N. (1979). 6-APA and the development of the β-lactam antibiotics. *J. Antimicrob. Chemother.* **5,** 7–14.

Rubens, C.E., McNiell, W.F. and Farrar, W.E. (1979). Evolution of multiple antibiotic resistance plasmids mediated by transposable plasmid DNA sequences. *J. Bacteriol.* **140,** 713–19.

Schollenberg, E. and Albritton, W.L. (1980). Antibiotic misuse in a pediatric teaching hospital. *Canadian Med. Assoc. J.* **122,** 49–52.

Shahrabadi, M.S., Bryan, L.E. and Van Den Elzen, H.M. (1975). Further properties of P-2 R-factors of *Pseudomonas aeruginosa* and their relationship to other plasmid groups. *Canadian J. Microbiol.* **21,** 592–605.

Stuy, J.H. (1980). Chromosomally integrated conjugative plasmids are common in antibiotic-resistant *Haemophilus influenzae. J. Bacteriol.* **142,** 925–30.

Sugarman, B. and Pesanti, E. (1980). Treatment failures secondary to *in vivo* development of drug resistance by microorganisms. *Reviews Infect. Dis.* **2,** 153–68.

Timoney, J.F. (1978). The epidemiology and genetics of antibiotic resistance of *Salmonella typhimurium* isolated from diseased animals in New York. *J. Infect. Dis.* **137,** 67–73.

Vastola, A.P., Altschaefl, J. and Harford, S. (1980). 5-epi-sisomicin and 5-epi-gentamicin B: substrates for aminoglycoside modifying enzymes that retain activity against aminoglycoside-resistant bacteria. *Antimicrob. Agents Chemother.* **17,** 799–802.

Weaver, S.S., Bodey, G.P. and LeBlanc, B.M. (1979). Thienamycin: New β-lactam antibiotic with potent broad spectrum activity. *Antimicrob. Agents Chemother.* **15,** 518–21.

Wise, R., Andrews, J.M. and Bedford, K.A. (1980). Clavulanic acid and CP-45,899: a comparison of their *in vitro* activity in combination with penicillins. *J. Antimicrob. Chemother.* **6,** 197–206.

Wise, R., Andrews, J.M. and Bedford, K.A. (1980). UK31214, a new aminoglycoside and derivative of kanamycin B. *Antimicrob. Agents Chemother.* **17,** 298–301.

APPENDIX OF ANTIBACTERIAL AGENTS

(A) Acting on cell wall turnover

Penicillins

Basal structure

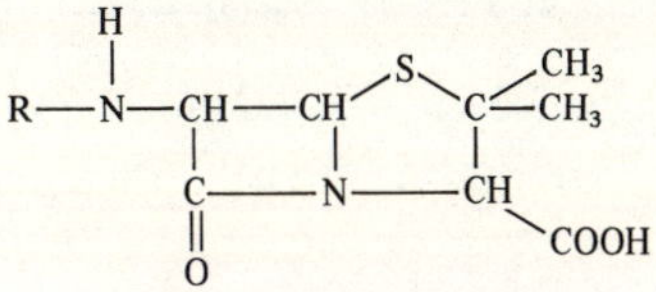

6-aminopenicillanic acid

Penicillin	*R-group*	*Comments*
penicillin G (benzylpenicillin)	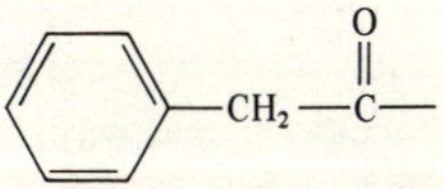	– acid labile – β-lactamase susceptible
penicillin V (phenoxymethyl-penicillin)	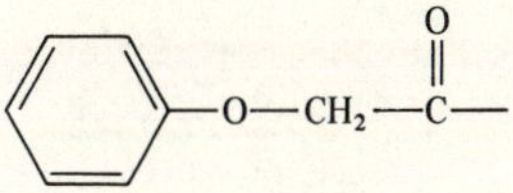	– acid resistant – β-lactamase susceptible
methicillin (dimethoxy-phenylpenicillin)	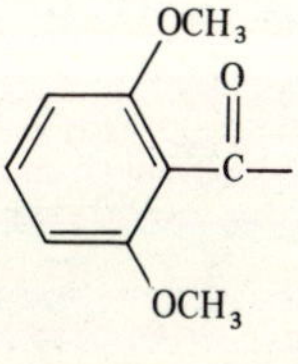	– acid labile – β-lactamase resistant – generally reduced activity relative to penicillin G
oxacillin (isoxazoyl penicillin)	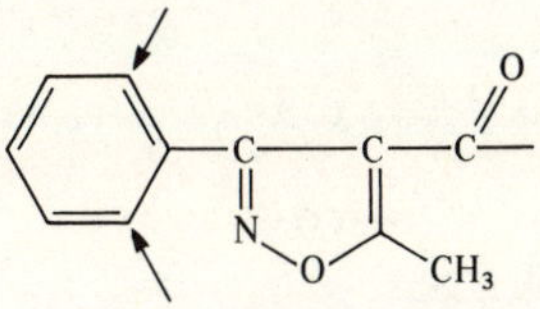 arrows – sites of halogen substitutions	– relatively acid stable – β-lactamase resistant – cloxacillin – one Cl – dicloxacillin – two Cl – flucloxacillin – one Cl, one F

Penicillin	*R-group*	*Comments*
nafcillin (ethoxynaptha-midopenicillin)	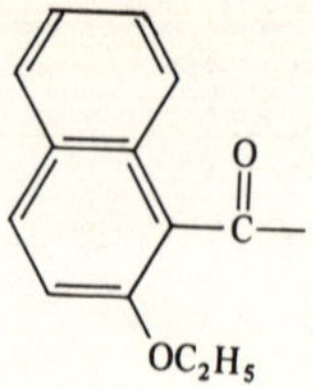	– β-lactamase resistant – like 'oxacillins' has higher antibacterial activity than methicillin – relatively acid stable
ampicillin ($\propto$-amino-benzylpenicillin)	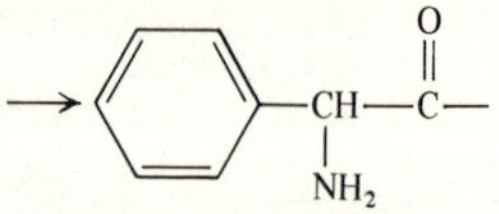 – arrow is site of OH in amoxycillin	– broader antibacterial spectrum than preceding penicillins – Staphylococcal β-lactamase susceptible – acid stable – amoxycillin better absorbed
carbenicillin ($\propto$-carboxy-benzylpenicillin)	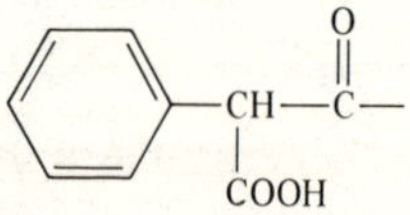	– broad spectrum – some anti-pseudomonal activity
ticarcillin	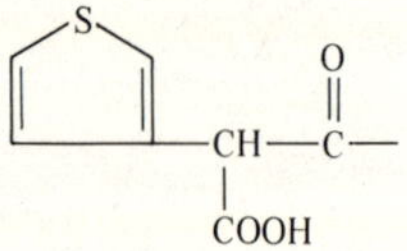	– more active anti-pseudomonal activity than carbenicillin
piperacillin	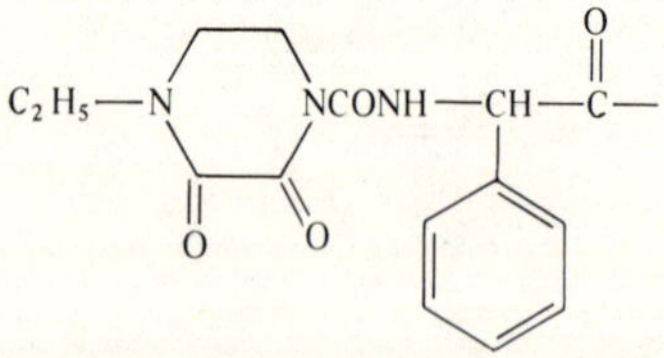	– broad spectrum – most active of group on *Pseudomonas*
azlocillin	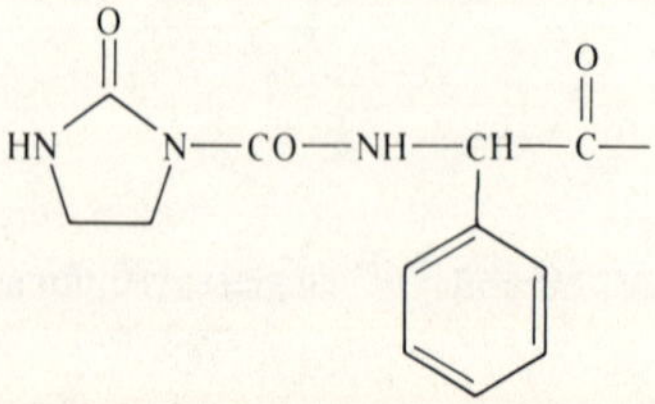	– broad spectrum – activity similar to piperacillin on *Pseudomonas*

Cephalosporins and cefamycins

Group 1 – β-lactamase susceptible and metabolically unstable

Basal structure

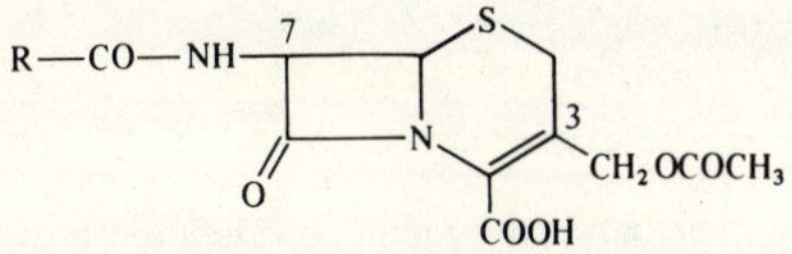

Compound	*R-group*	*Comments*
cephalothin	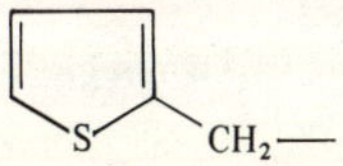	– deacetylated *in vivo* thus have short half-lives – broad spectrum – relatively low activity on *H. influenzae, N. gonorrhoeae,* very poor – *S. fecalis,* inactive – *P. aeruginosa,* indole – positive *Proteus* – also in this group are cephacetrile and cephaloram
cephapirin	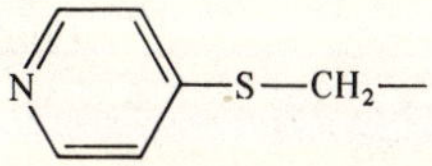	

Group 2 – β-lactamase susceptible, metabolically stable

Basal structure

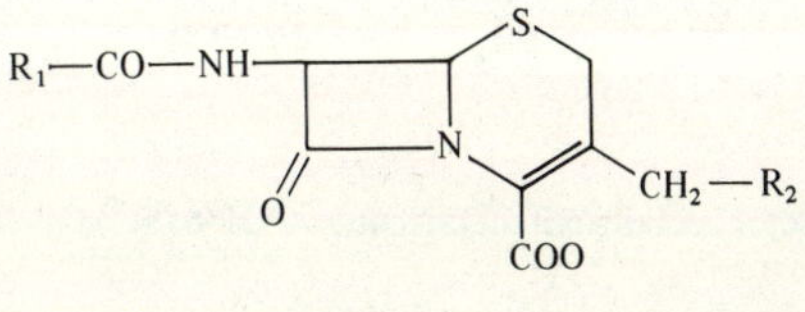

Compound	*R_1*	*R_2*	*Comments*
cephaloridine	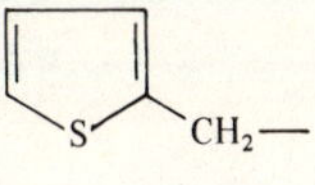	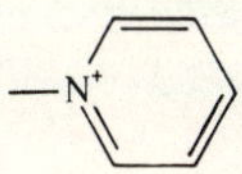	– similar anti-bacterial pattern to cephalothin – more nephrotoxic potential than cephalothin – higher serum levels than cephalothin

Compound	*R_1*	*R_2*	*Comments*
cefazolin	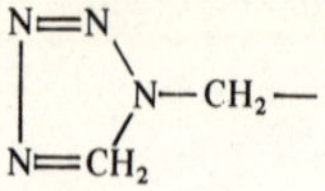	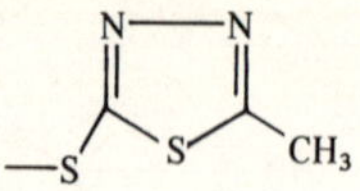	– cephaloridine acts on PBP.lb at concentrations below that affecting PBP3.

also in this group
ceftezole, cefazedone, cephanone, ceforanide, cefazaflur, cefotiam. Cefotiam has enhanced activity against gram-negative bacteria.

Group 3 – Compounds absorbed when given orally

Basal structure

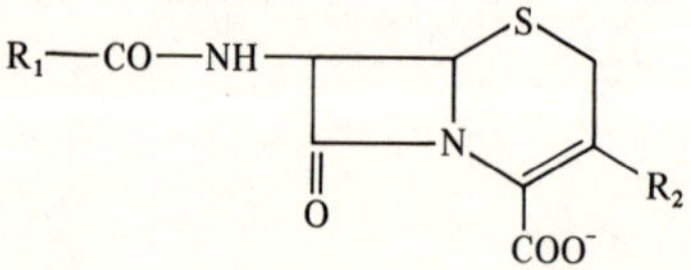

Compound	*R_1*	*R_2*	*Comments*
cephaglycin	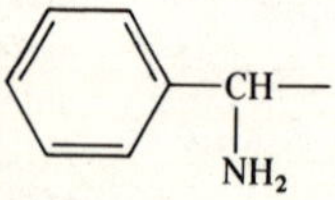	$-CH_2O-CO-CH_3$	– modest *β*-lactamase resistance – antibacterial spectrum similar to cephalothin – mainly active on PBP3 (filament production) – cephaglycin is deacetylated
cephradine	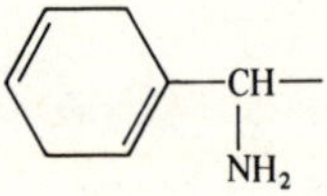	$-CH_3$	– clinical performance is probably similar

also in this group
cephalexin, cefaclor, cefadroxil, cefatrizine, CGP9000, SCE 100, FR 10612, RMI 19,592

Group 4 – parenteral cephalosporins with significant resistance to β-lactamases

Basal structure

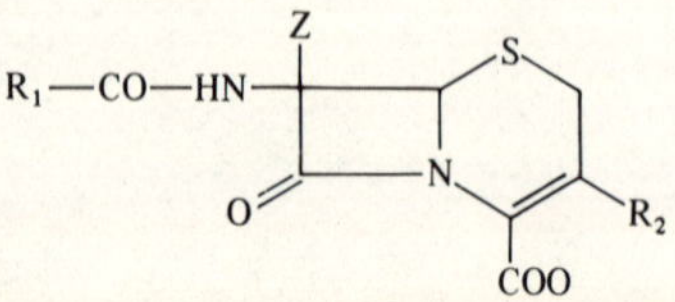

Compound	*R_1*	*R_2*	*Z*	*Comments*
(a)				
cefuroxime	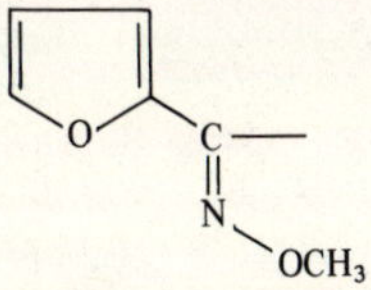	$—CH_2—OCONH_2$	H	– wide activity against many Enterobacteriaceae – no *Pseudomonas* activity – less active in general on *Bacteroides* than group 4d.
cefamandole	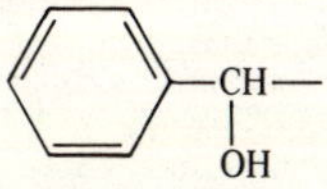	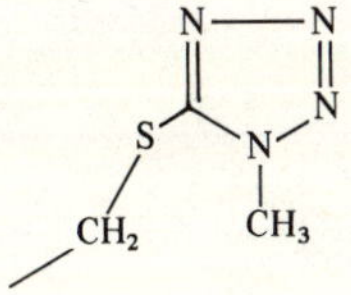	H	– good activity on *H. influenzae*. Cefuroxime very active on *N. gonorrhoeae*
also includes cefonicid				
(b)				
cefotaxime	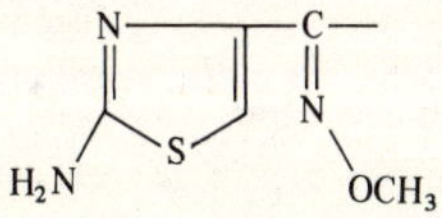	$—OCOCH_3$	H	– very active on Enterobacteriaceae. Some activity on *Pseudomonas* – cefotaxime – very active on *S. pyogenes, H. influenzae*
also includes SCE 1365 and FK 749				
(c)				
cefsulodin	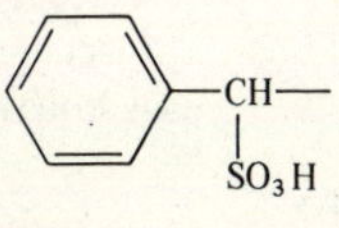	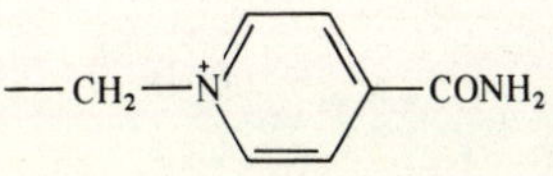	H	– cefsulodin, major potential is an antipseudomonal agent
cefoperazone	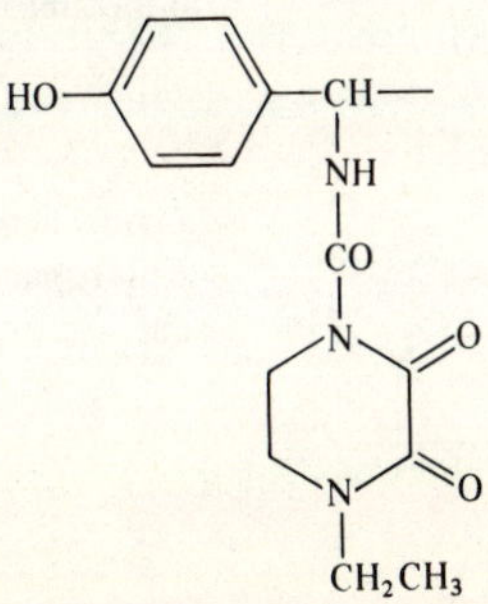	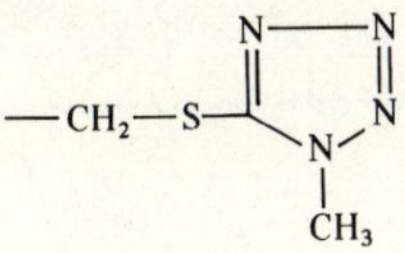	H	– wide spectrum of gram-negative bacteria including *P. aeruginosa* but not *E. aerogenes*

Compound	R_1	R_2	*Z*	*Comments*
(d) very resistant to *β*-lactamases				
cefoxitin	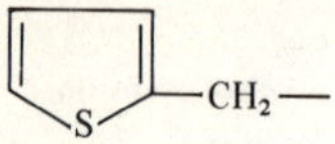	$—CH_2—O—CONH_2$	$—OCH_3$	– wide spectrum of gram-negative bacteria (but not *P. aeruginosa*) including *β*-lactamase producers. Activity less than moxalactam – good activity on *N. gonorrhoeae, B. fragilis*
moxalactam				– very wide spectrum on gram-negative bacilli, *H. influenzae, N. gonorrhoeae, B. fragilis*. Good activity on *Pseudomonas*

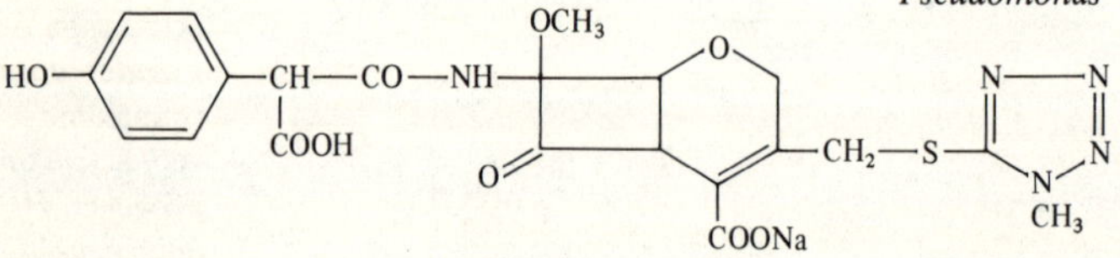

Thienamycin

– wide antibacterial spectrum
– inhibitor of *β*-lactamases and resistant to *β*-lactamase hydrolysis

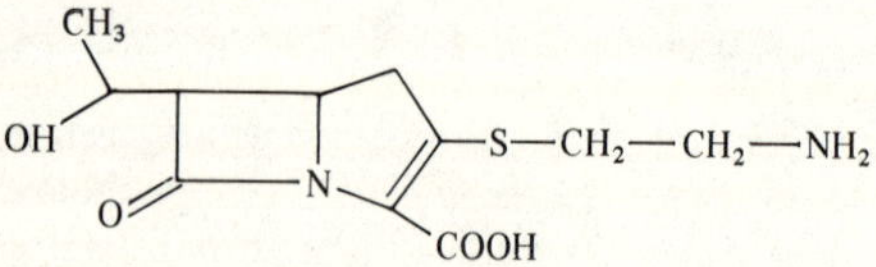

Clavulanic acid

– inhibits many *β*-lactamases

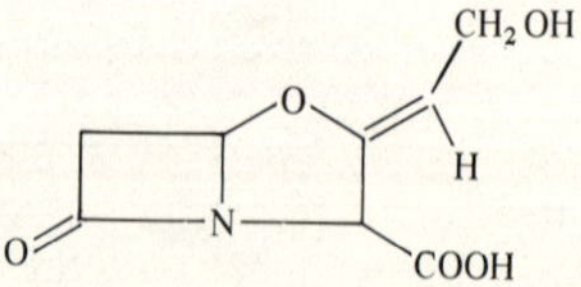

Mecillinam

– mainly active on Enterobacteriaceae

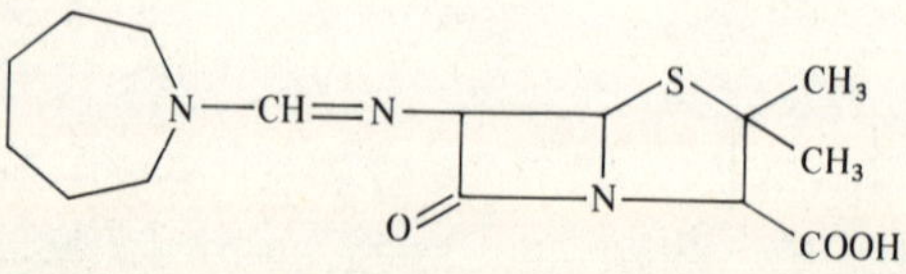

Compound — *Comments*

Phosphonomycin

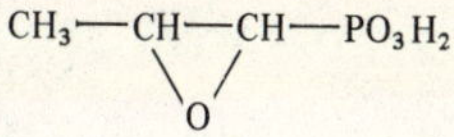

– analog of phosphoenolpyruvate

D-*cycloserine*

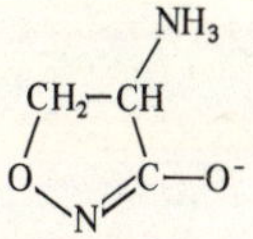

Fluoro-D-*alanine*

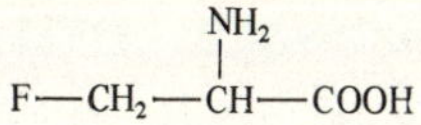

Bacitracin

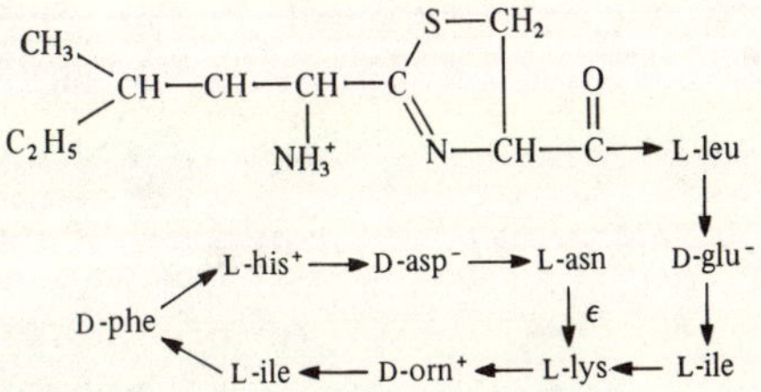

– leu, leucine; glu, glutamic; ile, isoleucine; lys, lysine; orn, ornithine; phe, phenylalanine; his, histidine; asp, aspartic; asn, asparagine

(B) Acting on cytoplasmic membrane

Polymyxins

DAB = ∝,*Y-diaminobutyric acid residue, MOA = (+)-6-methyl-octanoic acid residue, thr = threonine, phe = phenylalanine*

Polymyxin B_1

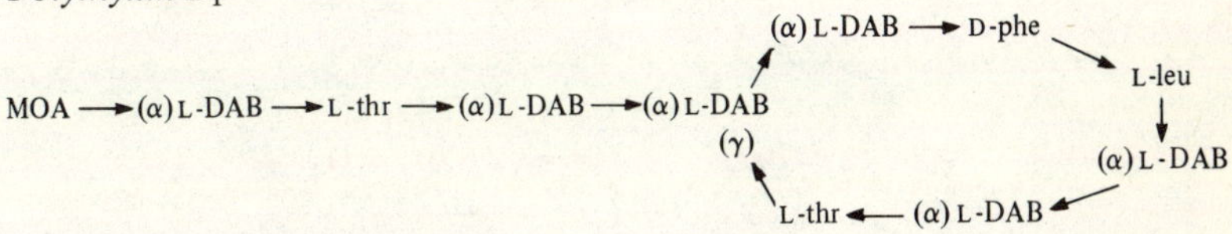

Polymyxin E_1 (colistin A)

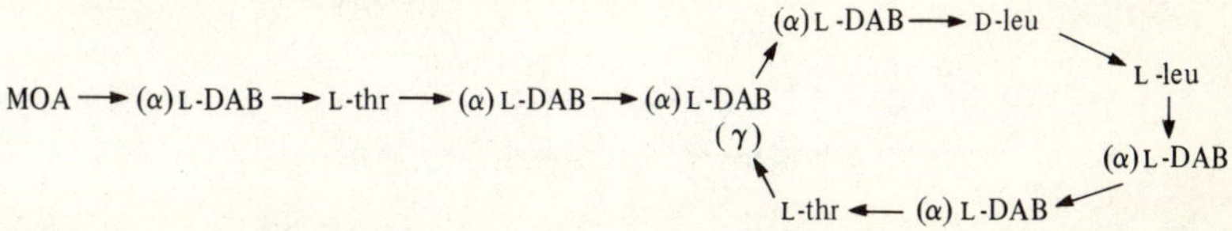

Amphotericin B (antifungal)

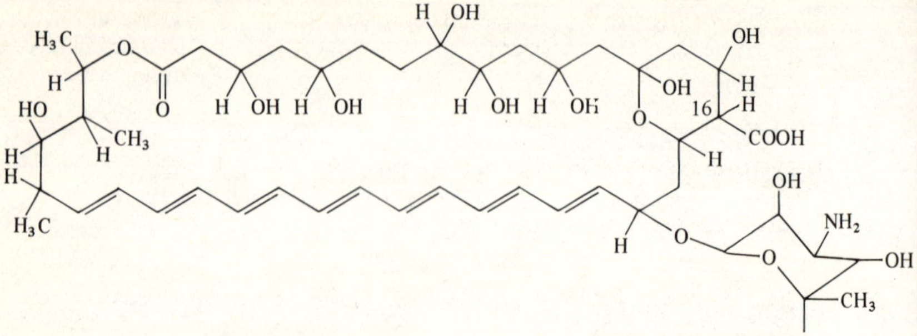

– amphotericin B methyl ester has a CH_3-COO-group at position 16 (methyl ester)

Chlorhexidine

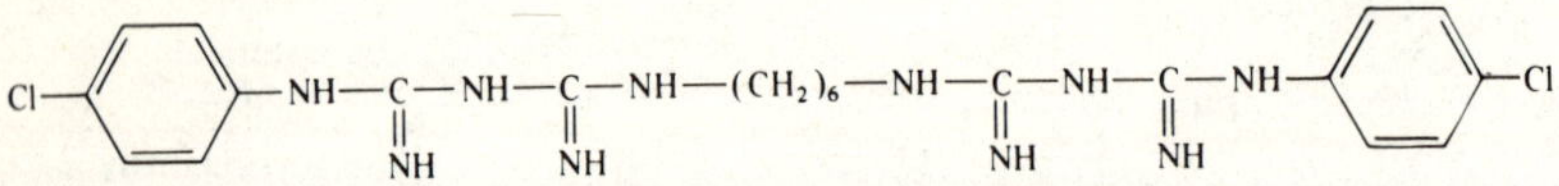

(C) Acting on nucleic acids or folate metabolism

Trimethoprim

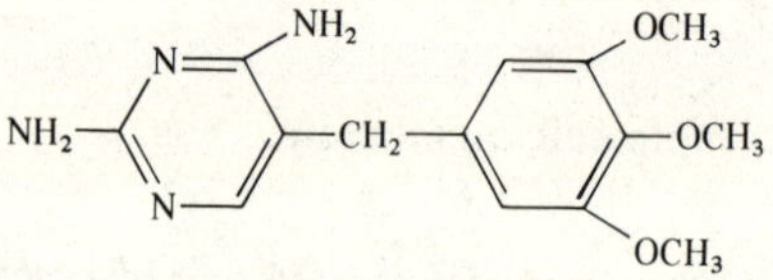

Pyrimethamine

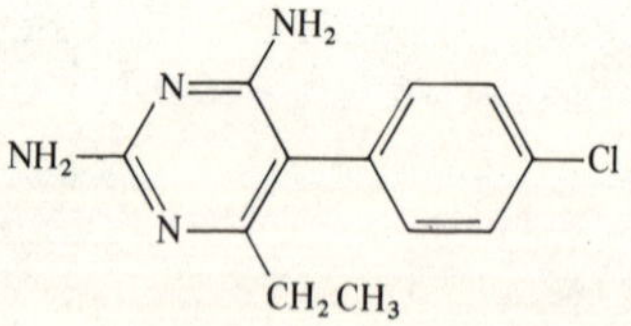

Sulfonamides

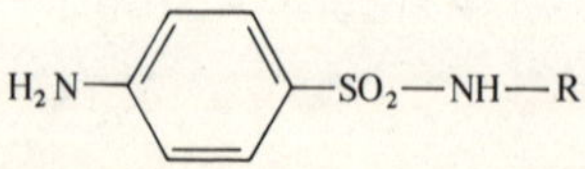

R-Group

sulfomethoxazole

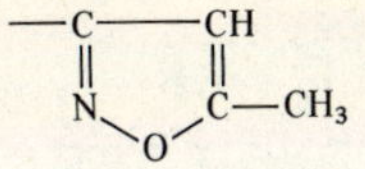

sulfisoxazole

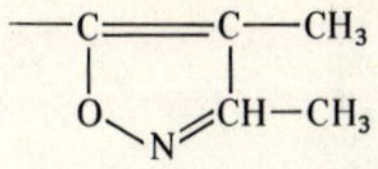

sulfadiazine

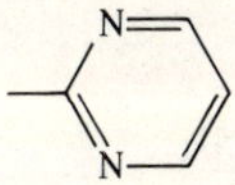

5-fluorocytosine

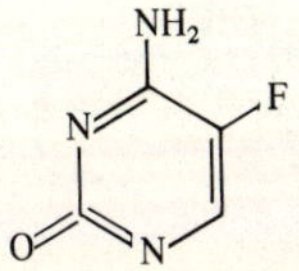

Adenine arabinoside (antiviral)

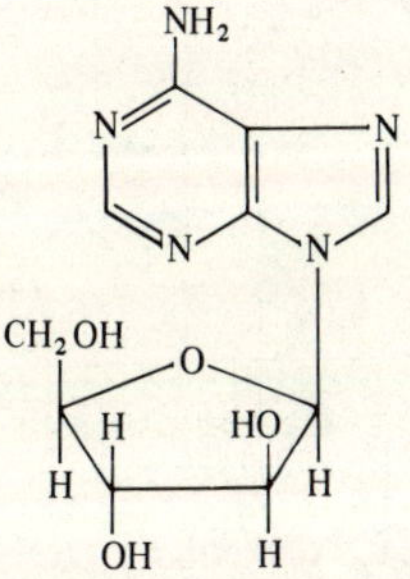

5-iododeoxyuridine (antiviral)

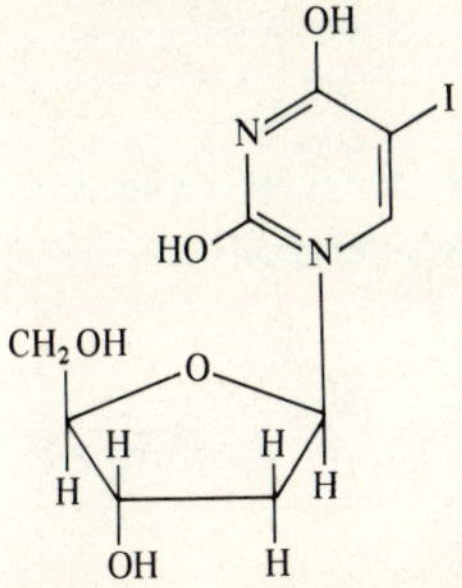

Chloroquine (antimalarial)

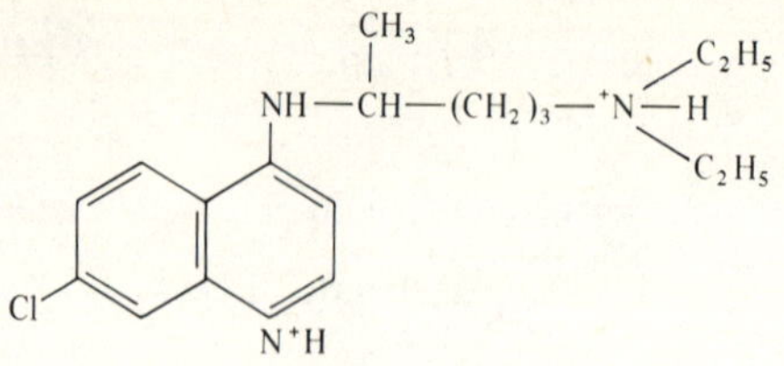

Rifampicin

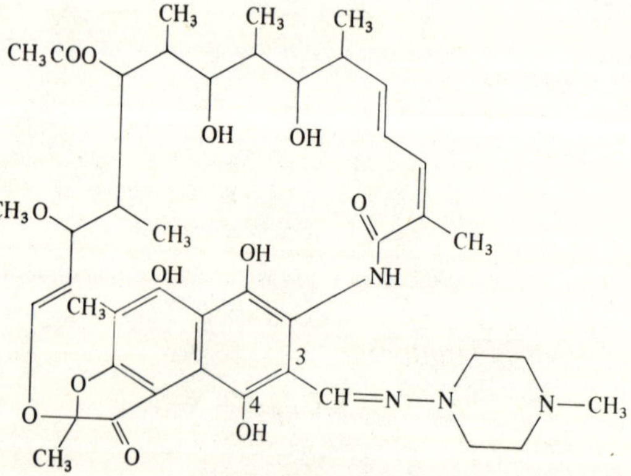

Nalidixic Acid

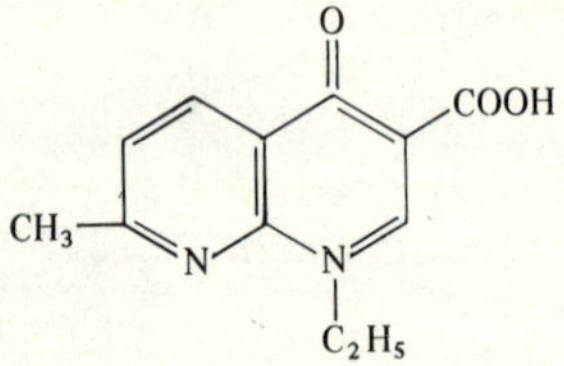

Novobiocin

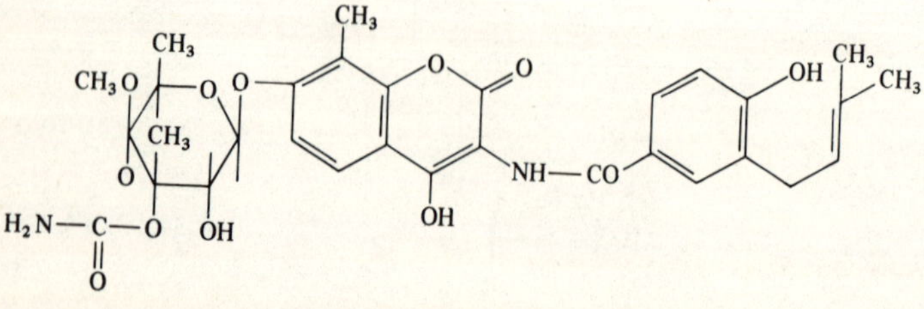

(D) Acting on protein synthesis

Chloramphenicol

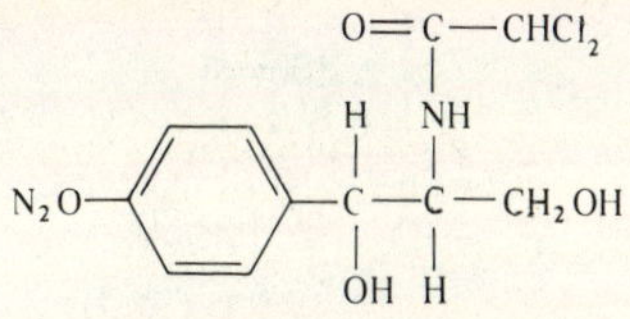

Puromycin

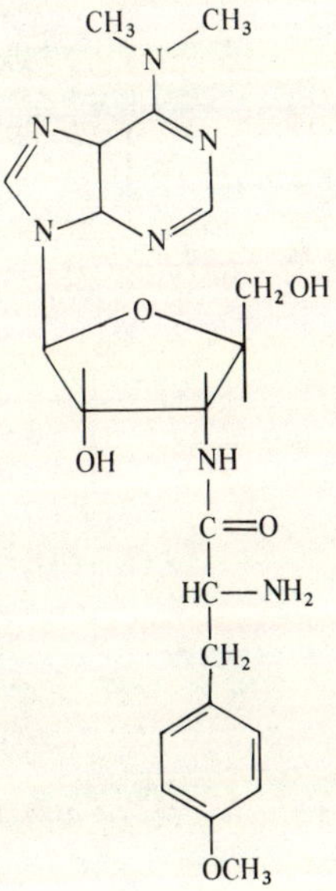

Erythromycin A

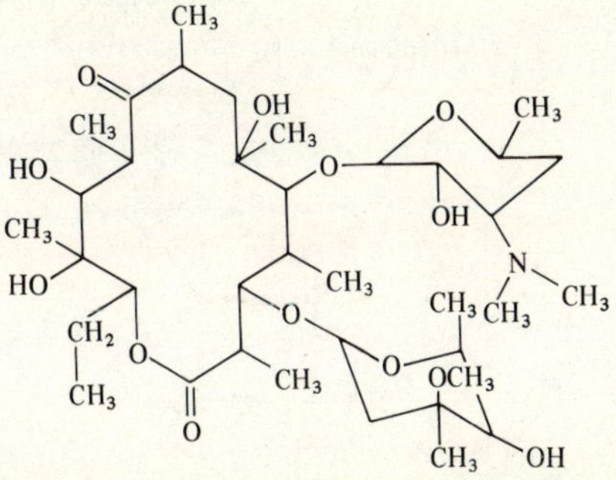

Lincomycin (clindamycin is 7-chloro-7-deoxy-lincomycin)

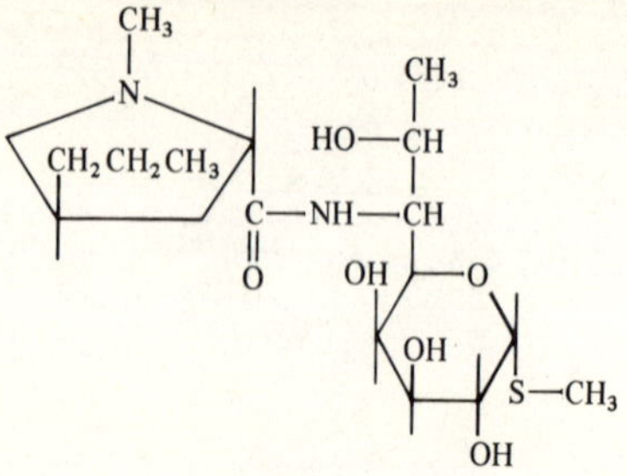

Tetracyclines

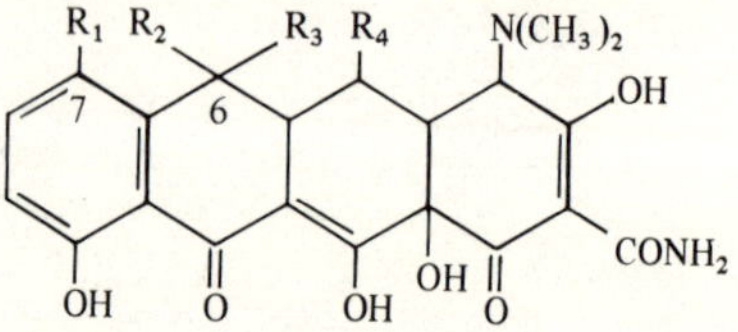

	R_1	R_2	R_3	R_4
tetracycline	H	CH_3	OH	H
oxytetracycline	H	CH_3	OH	OH
chlortetracycline	Cl	CH_3	OH	H
minocycline	$N(CH_3)_2$	H	H	H
doxycycline	H	CH_3	H	OH

Aminoglycosides (arrows show sites of attack of various modifying enzymes)

(a) Streptidine
streptomycin

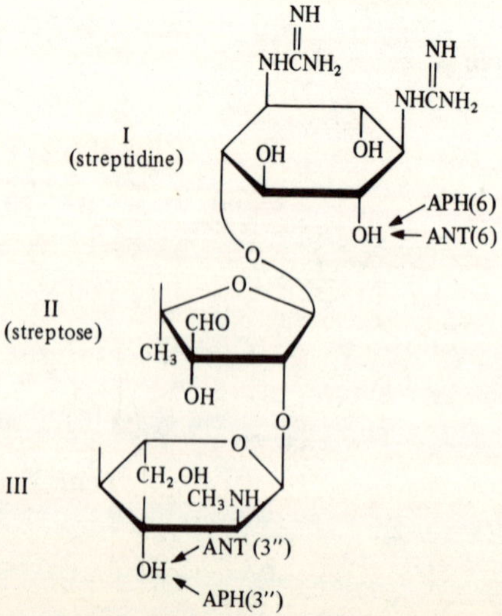

(b) Deoxystreptamine
kanamycin A

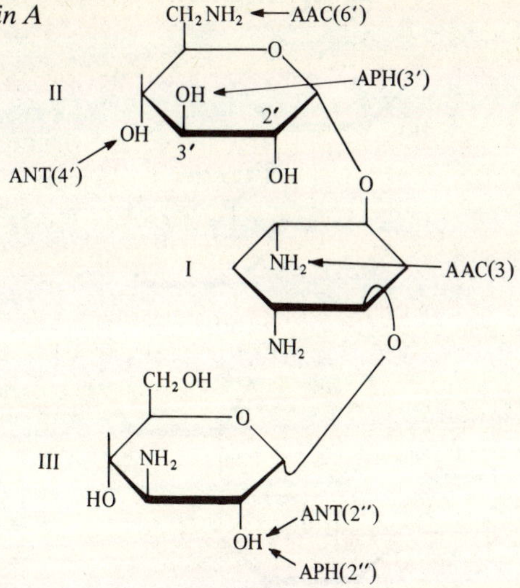

– amikacin is the 1-*N*-L(–)-γ-amino-∝-butyric acid derivative of kanamycin A
– kanamycin B has an amino group at the 2′ position (ring II)
– tobramycin is 3′ deoxykanamycin B

Gentamicins

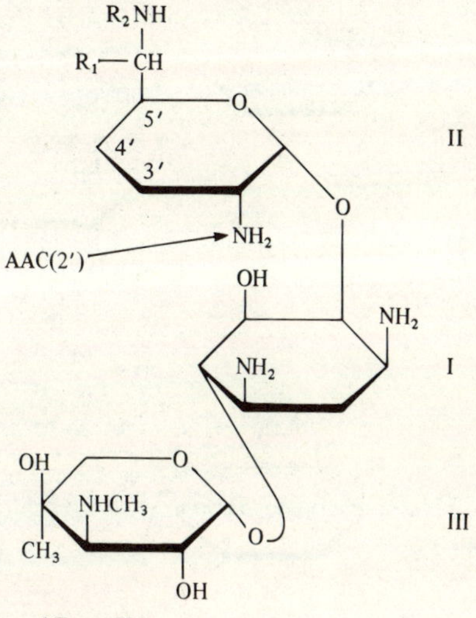

– gentamicin C_1, R_1 and $R_2 = CH_3$
gentamicin Cla, R_1 and $R_2 = H$
gentamicin C_2, $R_1 = CH_3$, $R_2 = H$
– sisomicin is 4′,5′ didehydrogentamicin Cla
– netilmicin is 1-*N*-ethylsisomicin

Neomycin B

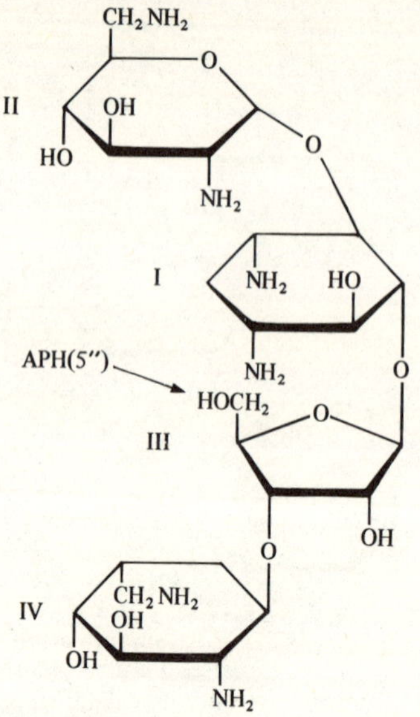

Kasugamycin

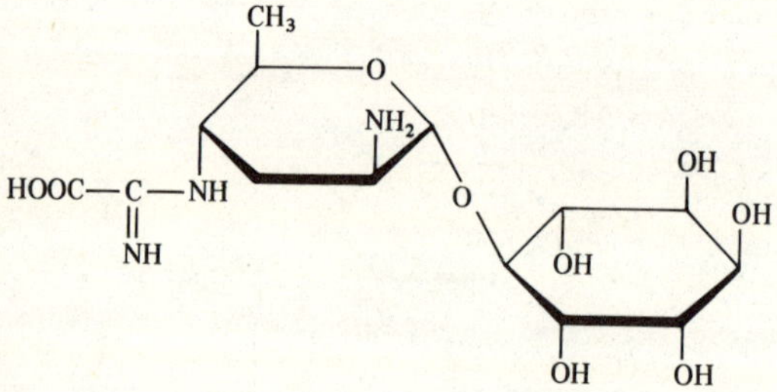

Aminocyclitols

spectinomycin

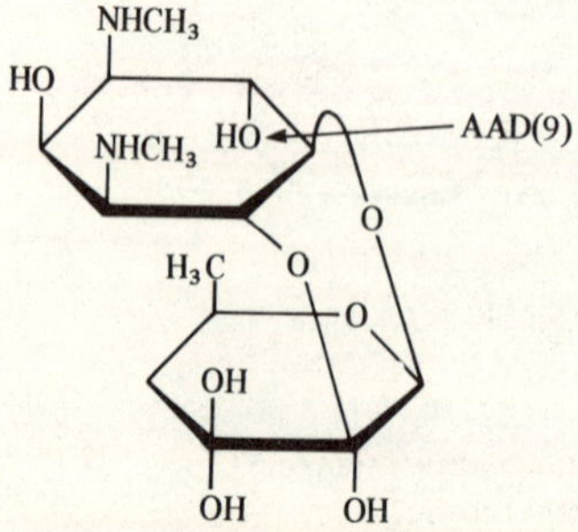

Fusidic acid

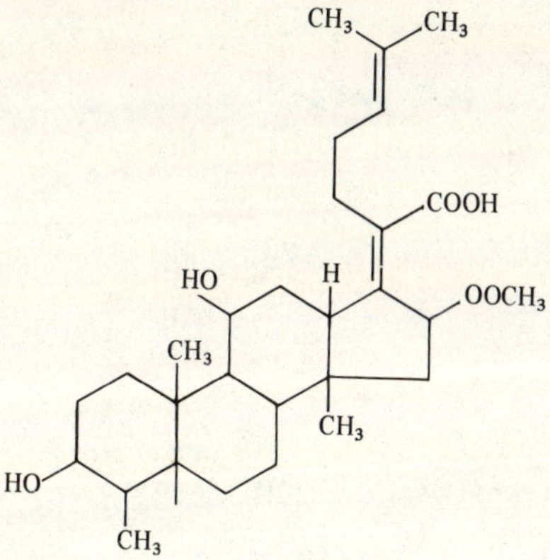

(E) Miscellaneous

Metronidazole

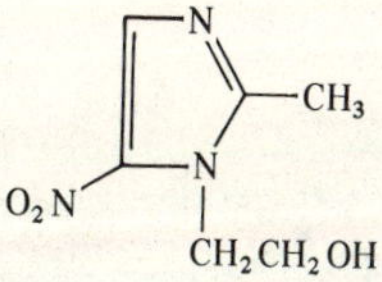

Nitrofurantoin

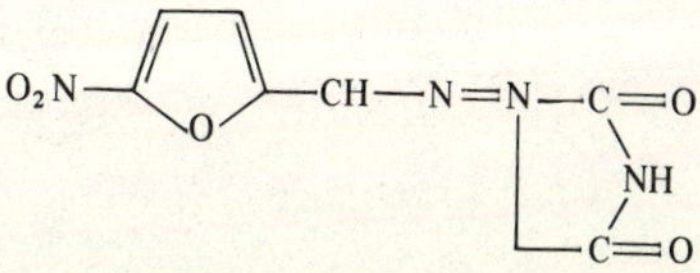

Isoniazid

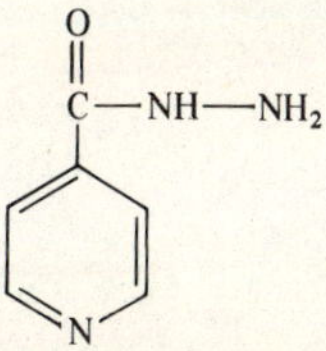

Ethambutol

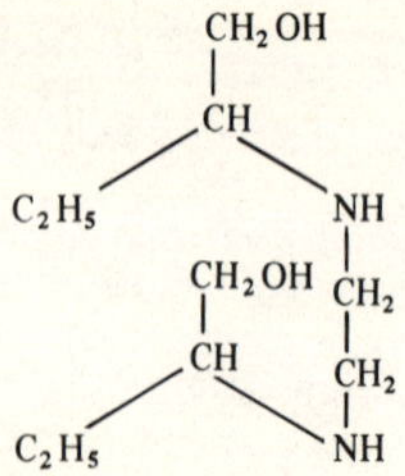

Ethionamide

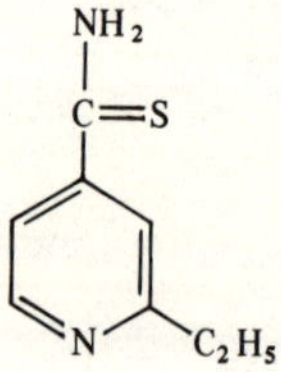

Pyrazinamide

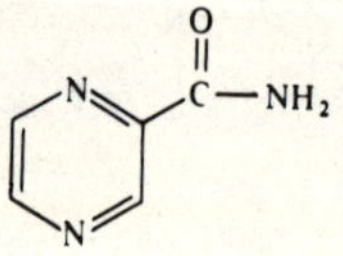

INDEX